Leigh Page

Stars of Earth

Wild Flowers of the Months

Leigh Page

Stars of Earth
Wild Flowers of the Months

ISBN/EAN: 9783337025816

Printed in Europe, USA, Canada, Australia, Japan

Cover: Foto ©berggeist007 / pixelio.de

More available books at **www.hansebooks.com**

OR,

WILD FLOWERS OF THE MONTHS.

BY

LEIGH PAGE.

EDINBURGH:

JOHNSTONE, HUNTER, AND CO.

MDCCCLXVIII.

MURRAY AND GIBB, EDINBURGH,
PRINTERS TO HER MAJESTY'S STATIONERY OFFICE.

'Spake full well, in language quaint and olden,
   One who dwelleth by the castled Rhine,
 When he called the flowers, so blue and golden,
   Stars, that in earth's firmament do shine ;—

Stars they are, wherein we read our history,
   As astrologers and seers of eld ;
Yet not wrapped about with awful mystery,
   Like the burning stars, which they beheld.

Wondrous truths, and manifold as wondrous,
   God hath written in those stars above ;
But not less in the bright flow'rets under us,
   Stands the revelation of His love.

Bright and glorious is that revelation,
   Written all over this great world of ours ;
Making evident our own creation,
   In these Stars of Earth,—these golden flowers.'

LONGFELLOW.

INTRODUCTION.

—◆—

'Flowers! flowers! bright, merry-faced flowers!
I bless ye in joyous, or saddened hours:
    I love ye dearly,
    Ye look so cheerly.
In summer, autumn, winter, or spring,
A flower is to me the loveliest thing
    That hath its birth
    On this chequered earth:
Oh! who will not chorus the lay I sing?'

WHO does not love flowers—fair, luxuriant wild-flowers—with which our earth is so beautiful? What pure, healthful thoughts they bring to the mind! with what warm, bright, happy feelings they stir the heart! They are 'a joy for ever.'

'Flowers are the brightest things which earth
    On her broad bosom loves to cherish;
Gay they appear, as children's mirth,
    Like fading dreams of hope they perish.

A

In every clime, in every age,
 Mankind have felt their pleasing sway ;
And lays to them have decked the page
 Of moralist and minstrel gay.'

Old and young, rich and poor, the learned and un-
taught, all acknowledge alike the common sentiment.
The humblest little child delights to gather them ; the
greatest scholar disdains not to bend his cultured intellect
in scanning their simplest beauties. They are bound in
our memories with the sports of childhood, the pleasures of youth. Who, in later years, can look upon even
the humble daisy, that raises its glad face from out the
meadow grass, or scent the fragrant hawthorn from the
roadside hedge, and not feel

 ' My childhood's earliest thoughts are link'd with thee :
 The sight of thee calls back the robin's song,
  Who, from the dark old tree
 Beside the door, sang clearly all day long,
  And I, secure in childish piety,
 Listen'd as if I heard an angel sing
 With news from heaven, which he did bring
 Fresh every day to my untainted ears,
 When birds, and flowers, and I were happy peers.'

Or who can see a group of village children roaming the
meadows in search of flowers, and not remember,

 ' I did the same in April time,
  And spoilt the daisy's earliest prime ;
  Robbed every primrose-root I met,
  And ofttimes got the root to set ;
  And joyful home each nosegay bore,
  And felt—as I shall feel no more ?'

Ah! it was with joy undimmed we wandered amongst the beauteous wild blossoms, ever ready with loving hearts to admire, wonder, linger over them; and with what force and eloquence they ever spoke to our hearts of affection and piety! Who can look upon a flower, and not feel the power, the wisdom, and beneficence of a Supreme Being, who decks our world with such heavenly beauties, who regulates the changing seasons, the bursting bud, the opening blossom, the ripening fruit, and falling leaf? All these teach us the truth of the Psalmist's words, that it is 'the *fool* who hath said in his heart, There is no God;' for

> 'There is a tongue in every leaf,
>   A voice in every rill—
> A voice that speaketh everywhere,
> In flood and fire, through earth and air,
>   A tongue that's never still.
>
> 'Tis the Great Spirit, wide diffused
>   Through everything we see,
> That with our spirits communeth
> Of things mysterious, life and death,
>   Time and eternity.'

I am sure, my dear young friends, *you* must all delight in flowers; and I purpose that we take each month a country ramble together, and see what loveliness adorns our earth, wandering through woods, and lanes, and sunny meadows, adown the streamlet's banks, or out upon the healthful, breezy moor, culling our little bouquet

as we go, and tracing God's love even in 'the flower of
the field ;' for

> 'There is a lesson in each flower,
>   A story in each stream and bower ;
>   In every herb on which you tread
>   Are written words, which, rightly read,
>   Will lead you from earth's fragrant sod
>   To hope, to holiness, and God.'

Therefore, in our rambles, by which we hope to add
to our enjoyment of nature, our further knowledge of
the useful and beautiful, and renewed vigour to mind
and body, let us be careful we 'rightly read' those
'written words ;'

> 'For, though we view each herb and flower
>   That sips the morning dew,
>   Did we not own Jehovah's power,
>   How vain were all we knew !'

I.

JANUARY.

ANUARY is a cold, withering month, and but few plants dare to expose the delicacy and beauty of their blossoms to its ungenial influence; yet still, in a clear, bright, frosty morning, we may venture forth, and hope to find a stray flower or two. We must not grieve if a carpet of snow hides from our eyes the beautiful grass-land; but re-member how beneath it lie, protected from nipping frost and chilling blast, the many roots and seeds that will spring and bud when summer skies appear. The trees are leafless; and one wonders where the sparrow or robin finds a sheltering nook; but their pretty twitting is still heard from joyous, grateful hearts, as they flit about the hedge-row, picking a chance hep from the dog-rose or white-thorn, and hiding, to enjoy their prize, beneath the foliage of a holly-bush.

Very beautiful is the holly (*Ilex aquifolium*) at all

5

seasons, but never more so than during these winter months, when its bright, dark leaves glance among the otherwise bare branches of the thorn, or privet hedge. In ancient days, this beautiful evergreen was called Holy

HOLLY—*Ilex aquifolium.*

Tree, on account of its being used to deck the churches at Christmas time, and hence the corruption of *holly.* Its pretty, white, wax-like flower does not appear until April; but is scarcely so beautiful as its poisonous scarlet berry, which glows so brightly in this winter month. The glossy leaves are usually edged with sharp spines, but those on the upper branches are often smooth and harmless, as Southey describes in his beautiful little poem on the holly-tree:

'Below a circling fence its leaves are seen,
    Wrinkled and keen ;
No grazing cattle, through their prickly round,
    Can reach to wound ;
But, as they grow where nothing is to fear,
Smooth and unarmed the pointless leaves appear.'

The holly-wood is very hard and white, and is much used by the turner and mathematical instrument maker. The bark contains a substance from which bird-lime is made.

It may be that the snow has melted from yon gently rising hillock that faces the sun, and we may find a sprig or two of common chickweed or stitchwort (*Stellaria media*) to carry home to our pet canary. It is a pretty, elegant plant, with pearly-white starry blossoms, delicate green leaves, and spreading, branched, brittle stems. It flowers throughout the entire year, and its constantly ripening seeds form an abundant supply of food to the birds.

'The little chickweed, oh, the little chickweed !
'Tis a simple flower, and is sweet indeed—
Sweet and ingenuous as a child,
Running about in the woodlands wild.'

There are eight species of stitchwort, but perhaps the

prettiest of all is the greater stitchwort (*Stellaria holostea*), often called satin-flower, which shows its delicate cluster-ing blossoms in the hedgeside during May. It is a more erect plant than the common chickweed, its slender stems often growing as high as two feet. The lesser stitchwort (*Stellaria graminea*) is a still more weak and brittle plant, its stems growing longer, and supporting them-selves amongst the surrounding bushes, where it loves best to bloom. The Alpine stitchwort (*Stellaria ceras-toides*) is a low plant, with fine silky leaves, common, I believe, in the Highlands of Scotland.

The Latin name of *Stella* signifies *star*, and is there-fore very appropriate to our pretty, silvery, star-like blossom. Its name of stitchwort is derived from its sup-posed usefulness in curing pains or *stitches* in the chest or side ; and that of chickweed, because hens feeding upon it were believed to lay more eggs.

As we wander onwards, we shall doubtless find a flaunting dandelion (*Leontodon taraxacum*), which is familiar to almost every eye. Common, and generally despised, it is nevertheless a handsome flower, glittering like a golden star in the fields and meadows of the new-born year. Its leaves, which are large and coarse, grow from the root; and, being deeply cut, with the lobes turned backward, its name was consequently made from two Greek words, signifying *lion* and *tooth*. Its root is a medicine used in England ; its leaves are sold in the French markets for salad ; and in some countries the

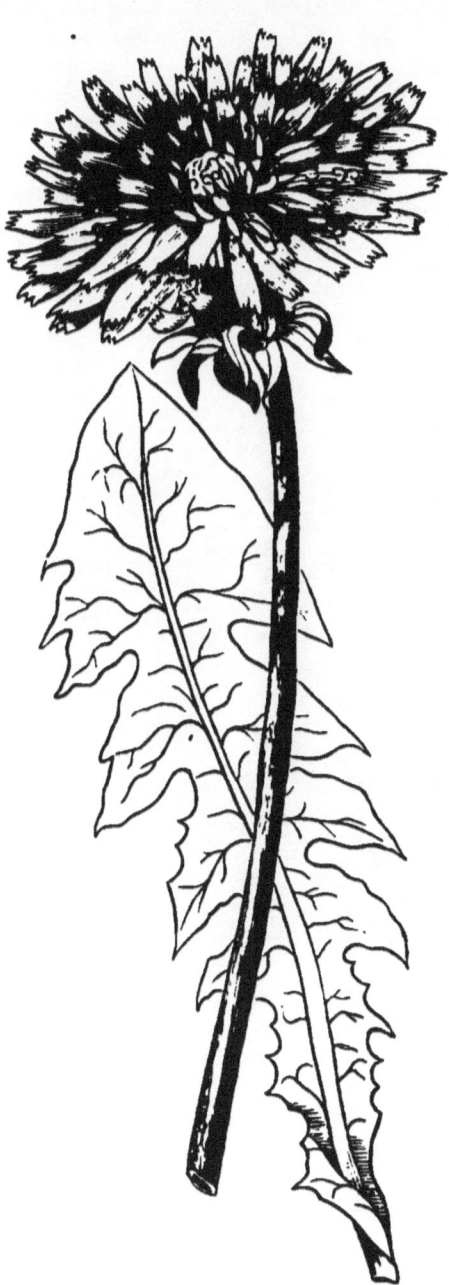

DANDELION—*Leontodon tarazacum.*

young roots are roasted and ground, and mixed with
coffee. The stems are full of a milky juice, which
indelibly stains linen, or the fingers which gather them.
Every one knows its downy balls of seed, that succeed
the flower; how, as children, we called them 'clocks,'
and blew, and blew, to ascertain the fleeting hour of the
summer's day.

> 'The dandelion, with globe of down,
> The schoolboy's clock in every town,
> Which the truant puffs amain,
> To conjure lost hours back again.'

When Linnæus, the great Swedish botanist, proposed
the use of a floral clock, to be composed of plants which
opened and closed their blossoms at particular hours, the
dandelion was one he selected on account of its petals
opening at six in the morning. The hawkweed was
another—it unfolds at seven, the succory at eight, the
celandine at nine, and so on; the closing of the blossoms
marking the corresponding hours of afternoon and evening. Nor is this the specific effect of light and air, as
they are found to do the same if placed in a dark room.

> ''Twas a lovely thought, to mark the hours,
>   As they floated in light away,
> By the opening and the folding flowers,
>   That laugh to the summer's day.
>
> Thus had each hour its own rich hue,
>   And its graceful cup or bell,
> In whose coloured vase might sleep the dew,
>   Like a pearl in an ocean shell.

> Oh ! let us live, so that flower by flower
>   Shutting in turn, may leave
> A lingerer still for the sunset hour,
>   A charm for the shaded eve.'

But now we have reached our little sunny bank, the snow has disappeared from the grass, and see what is raising its glad face to the genial sunshine !

> 'A daisy, with its pinky lashes.'

Pluck it, and look at its delicate white petals, surrounding a golden centre, and backed by a delicate star of green !

> 'Wee, modest, crimson-tipped flower,'

What poet has not loved to sing of it ! 'The robin of flowers ;'

> 'The golden tufte within a silver crowne ;'

'The poet's darling,' as Wordsworth lovingly designates it. Ben Jonson and Chaucer call it 'Day's Eye,' because it is

> 'The little daisy, that at evening closes.'

The Scotch give it the pretty title of gowan—

> 'The opening gowan, wet wi' dew.'

In Italy it bears the name of meadow-flower ; in France, that of marguerite. It was the device of the unfortunate Marguerite of Anjou, and also Marguerite of Valois, that noble, pious princess, the protecting friend of Calvin, so famous for her wit and beauty, so beloved by all around her, who was called by her brother, Francis

CHRISTMAS ROSE—*Helleborus niger.*

the First, king of France, his 'Marguerite of Marguerites.'
Its botanical name is *Bellis perennis*, derived from the
Latin *bellus*, beautiful.

And now, with our sprig of holly, our chickweed, dandelion, and bunch of daisies, let us wander from those
sunny fields to the open heath-land beyond, and gather
a few golden flowers from 'the never bloomless furze,'
which loves a sandy soil, and is to be found so plentifully
on high lands.

> ' Fringing the fence on sandy wold
> With blaze of vegetable gold,
> The furze (but ah ! beware the thorn,
> Too oft 'mid brightest blossoms born !)—
> The furze still yields its fragrant bloom.'

The dwarf species (*Ulex nanus*) is a low, prickly shrub,
its golden blossoms opening first in autumn, but lingering on through the winter months, defying winter frosts,
and gleaming in beauty when stormy blasts have withered
every other flower. The common gorse, or whin (*Ulex
Europæus*), is a very bushy shrub, growing from two to
five feet high, beset with thorns, and its numerous papilionaceous, or butterfly-like blossoms, often clothing the
hill side with a lustrous beauty. It does not continue to
bloom quite so late as the dwarf species ; but as early as
May the moor will gleam like a sheet of gold under the
sunlight, a glorious mass of yellow blossoms ; and most
delicious is the odour of its fragrance, borne on the soft
summer air. When Linnæus first beheld our native

moorland in all its loveliness, he is said to have fallen on his knees, and thanked God for the sight.

Small is our January bouquet, and yet with such must we be content. In the garden we may find a Christmas rose (*Helleborus niger*) nestling in its bed of dark, shining leaves. It is a poisonous plant, as are also the wild varieties, which, however, do not blossom before February, and have greenish purple flowers. The ancients held the Christmas rose in great esteem and veneration, and used in winter time to scatter their floors with its beautiful white blossoms, to hallow their houses, and guard themselves from the power of evil spirits. Let us gather one of its short, thick-stemmed flowers, and, placing its snowy whiteness in contrast with yellow furze and glistening holly, deck the parlour table with this scanty show of winter blossoms. Then let us circle around the comfortable, cheery hearthstone, to discuss our pleasant walk ; and, dwelling on all the wonders and beauties which it has revealed, let us pray :

> ‘ Instruct us, Lord,
> E’en by the simple lesson of a flower,
> To live by Thee.’

II.

FEBRUARY.

———•———

'Come forth, ye lovely heralds of the spring!
　Leave, at your Maker's call, your earthly bed ;
At his behest your grateful tribute bring :
　To light and life, from darkness and the dead,
　Thou timid snowdrop, lift thy lowly head.'

HE cold ungenial month of February has arrived, with its chilling winds and sleety showers, its alternate changes of snow and thaw; but even in spite of these we may find a bright sunny morning to tempt us out into the sheltered lane, and be able to gather a small nosegay out of the few wild-flowers that dare—at the season when the Almighty 'scattereth his hoar-frost like ashes'—to brave the wintry blast, and proclaim, in their exquisite perfection and beauty,

　'The hand that made us is divine.'

Many of the trees and bushes around us are beginning

to give some faint indications of existence, to remind us the beautiful spring-time is close at hand, when

> ‘ The folded leaf is woo’d from out the bud.’

The fresh green of the elder-tree and gooseberry, the darker leaf of the honeysuckle, or perchance an early ‘catkin,’ or tassel-like blossom of the hazel, may be seen; the faint aromatic odour of the ground ivy detected, though it does not flower until a month later. Its young leaves, however, are often gathered now by country people, as a cure for coughs and colds. Early as it is, we may chance to find, on some rubbish heap near the garden, an early blossom of the dingy hellebores, of which we have two species: the green hellebore (*Helleborus viridus*); and the stinking hellebore (*Helleborus fœtidus*). Both are highly poisonous, blooming in woods and waste places, and bearing greenish flowers, those of the latter somewhat larger and tinged with purple. The green leaves are numerous and pedate, or *bird-footed;* the stems are usually short, thick, and sturdy.

At this time, also, the bright little winter aconite (*Eranthis hyemalis*) raises its yellow satiny cup above the ground:

> ‘ The aconite that decks with gold
> Its merry little face.’

At one time it belonged to the hellebore tribe, but is now classed among the ranunculus family. Its green leaves spread like a drooping frill around the neck of the stem, close beneath the flower. It grows near to the ground, and from this has derived its botanical name from the Greek, signifying *earth* flower. Like all the aconites, it possesses highly poisonous properties. It is a beauteous little blossom, which we value all the more for appearing at this dreary season, when so few flowers gladden our sight. Well has Clare described it,

> ' The winter aconite,
> With buttercup-like flowers, that shut at night,
> Its green leaf furling round its cup of gold,
> Like tender maiden muffled from the cold.'

But—

> ' Already now the snowdrop dares appear,
> The first pale blossom of the unripened year ;
> As Flora's breath, by some transforming power,
> Had changed an icicle into a flower.'

Let us hunt first for this exquisite herald of spring, 'the fair maid of February,' as it has been called, which, though considered a doubtful wildling, may often be found adorning the lanes of the south of England. With the earliest gleam of sunshine, when nature still rests in her winter's sleep, and only the robin is heard piping his little song of gladness, then will

> ' The snowdrop, who, in habit white and plain,
> Comes on, the herald of fair Flora's train,'

venture to peep from the compact little flower-sheath, in which it has so long lain snugly enveloped, and show its

beauteous fair face above the snow, which it rivals in whiteness :

    'Lone flower, hemmed in with snows, and white as they.'

Who does not experience a gush of true pleasure in

SNOWDROP—*Galanthus.*

discovering the first snowdrop ? or fail to welcome with
gladness, the little bunches of white blossoms that peep
from out their nest beneath the sheltering tree, bringing
to the flowerless earth the cheerful promise of brighter
skies, and more genial sunshine ?

> ' Like a star on winter's brow,
>   Or a gleam of consolation
>   In the midst of sorrow, thou
>   Comest, pearl of vegetation.'

How beautiful are its pair of straight-veined, pale-green leaves ; its drooping bells, composed of three outer petals, enclosing their inner triplet delicately edged with green ! How daintily they bend from the tender stalks,

> 'Like pendent flakes of vegetating snow,'

braving the chilling frosts, yet timidly gathering in clusters, nodding and trembling 'neath every sweeping blast ! Exquisite, delicate little blossom, how we love thee ! coming in this wintry season, when so few others visit us. Its botanical name (*Galanthus*) is taken from the Greek, and signifies milk-flower. The Germans call it snow-bell ; the French give it the name of *perceniege*—snow-piercer, which is even prettier than our own.

The snowdrop is considered the emblem of hope, doubtless because it comes

> ' The early herald of the infant year ;'

but a pretty legend tells us of its first adoption, which my young friends may be amused to hear.

It is said that Hope, one day standing leaning on her anchor, watching the snow as it fell on the earth, and Spring patiently waiting till the wintry blast had passed away, lamented that those beautiful white flakes were not fair blossoms to gladden the land, rather than chilling snow to leave all so bare and desolate. Upon which Spring extended her fair arms, poured forth her sweetest smiles and warmest breath on the falling flakes, which immediately assumed the form of flowers, and dropped on the earth in scattered clusters of beauteous snowdrops. Hope, enchanted with the sight, caught the first trembling blossom as it fell, and at once adopted it as her emblem. Thus

> ‘ The pallid snowdrop, like the bow
>   That spans the cloudy sky,
> Comes fraught with hope, for well we know
>   That brighter days are nigh ;
> That circling seasons, in a race
> That knows no lagging, lingering pace,
> Shall each the other nimbly chase,
> Till Time’s departing final day
> Sweep snowdrops and the world away !’

Another doubtful wildling to be gathered now, is the spring crocus (*Crocus vernus*), which so often blooms in our garden borders, and is frequently to be found in little tufts adorning the pastures of the southern counties of England. Its flowers are lilac, and, like the snowdrop, it possesses erect, narrow, green leaves, and a bulbous root, which all autumn and winter lies hidden beneath the ground, but seems to leap into life at the first summons of spring.

> ‘ Down in its solitude under the snow,
>   Where nothing cheering can reach it ;

There, without light to see how to grow,
    It trusts to kind nature to teach it.

It does not despair, nor be idle, nor frown,
    Locked in so gloomy a dwelling;
Its leaves they run up, and its roots they run down,
    Whilst the bud in its bosom is swelling.'

The crocus is said to be named after a beautiful youth, who was changed into a flower by the gods, on account of his impatient love for a shepherdess named *Slimax*, who, in consequence, died through grief, and was turned into a yew-tree.

But see! we have come upon the pretty, small, blue blossoms of the procumbent speedwell (*Veronica agrestis*), which trails its stems along the ground, and is the first of all the speedwells. We have eighteen kinds; but the largest, and perhaps the best known of all,

'Brighter than bright heaven the speedwell
    blue,'

is the germander speedwell (*Veronica chamædrys*), which decks with its brilliant blue blossoms the April

CROCUS—*Crocus vernus.*

meadows, and which we shall gather in that month.

On yonder sheltered hedgebank we shall likely find the leafy blossom of the red dead nettle (*Lamium purpureum*), which is common by every wayside, and blooms throughout the summer. Its dark-green leaves faintly tinged with purple, and its purplish-red flowers, are not beautiful, neither is its smell agreeable; but we prize it now when our nosegay is so small. Its foliage is somewhat like the stinging nettle, which we are all afraid to touch; but it is called '*dead*,' or '*blind*,' because possessing no venomous power.

Surely a gentle perfume scents the air; and an adventurous little violet (*Viola odorata*) is to be found in some sheltered nook, half hidden by its heart-shaped green leaves, but betrayed by its own sweet odour.

> 'In the woods this simple flower
> Conceals its purple crest,
> But from out her grassy bower,
> Her scent betrays her nest.'

Oh, who will be first to pluck it from its retreat, with that vague happy feeling that theirs is mayhap the first eye that has lighted on it since it was fashioned by the hand of God! All poets love the

> 'Violets dim,
> But sweeter than the lids of Juno's eyes,
> Or Cytherea's breath.'

Sir Walter Scott tells us:

> 'The violet in her greenwood bower,
> Where birchen boughs with hazel mingle,

> May boast itself the fairest flower,
>   In glen, in copse, or forest dingle.'

But the shrinking violet is too fit an emblem of modesty to boast. An old English sonnet speaks of it also as the emblem of faithfulness. It is a native of every part of Europe; and many a copse-wood in England is made beautiful by its deep purple, and sometimes white blossoms, so rich and generous in their fragrant sweetness. Our greatest of English poets has declared it to be a

> 'Wasteful and ridiculous excess
>   To throw a perfume on the violet.'

The dog-violet (*Viola canina*) is a larger, brighter flower, holding its head erect to view. Its blossoms are beautifully streaked with purple lines, but scentless. It is of it that Clare has so exquisitely written :

> 'Violets, sweet tenants of the shade,
>   In purple's richest pride arrayed,
>   Your errand here fulfil :
>   Go, bid the artist's simple stain
>   Your lustre imitate in vain,
>   And match your Maker's skill.'

There are several other species; but rarely are any found to bloom before March or April. So gather our precious early-comer gently, for its delicate flower is fragile, and soon fades.

> 'The trembling violet, which eyes
>   The sun but once, and, unrepining, dies !'

The marsh violet (*Viola palustris*) bears a pretty, transparent-looking flower. It is an inhabitant of bogs, and very plentiful in Scotland; as is also the yellow violet, or mountain pansy (*Viola lutea*), that decks the moorland heaths. A syrup is made from violets, which is most useful in chemistry.

We must not forget to look for a bit of common groundsel (*Senecio vulgaris*) as we wander along, which provides such ample store of food for the little birds that gladden our woodland walk with the varied notes of their full-hearted song. At almost all seasons of the year, in summer's warmth or winter's cold, we may find it springing up, a true emblem of constancy:

> ' Through storm and wind,
> Sunshine and shower,
> Still will you find
> Groundsel in flower.'

There are, I believe, nearly six hundred species of groundsel, and nine are British kinds. It is to be found in almost every British settlement; its seeds having doubtless been conveyed in the grain taken to foreign lands, as has happened with many of our wild-flowers.

The spurge laurel (*Daphne laureola*), with its bright evergreen leaves, has been conspicuous during all the winter months; but now we chance to see its pale yellowish-green pendent flowers, which hang in clusters, like little waxen bells, beneath its dark, shining, palm-like leaves, situated in circular rows around the extremity of its few branching stems, which grow often three feet high. Unlike the *Daphne mezereon*, which we shall gather next month, it is not sweet-scented, though its bluish-black berries are equally poisonous to all but birds, and eagerly sought for by the thrush and black-bird. The spurge laurel is a native of Britain, more common than the mezereon, and may in mild seasons be gathered as early as January. It is one of the hand-somest evergreens in the woods and hedges of the south of England, but is more rare in Scotland, and gleams as brightly during winter's storms as 'neath the summer sunshine. Its fruit is highly useful as a medicine. The genus is named from the nymph so beloved of Apollo, who, imploring the protection of the gods, was changed by them into a laurel.

The sweet fragrant blossoms of the common rosemary (*Rosmarinus officinalis*) now also begin to show on the low shrubby evergreen bush, and very pretty are its pale blue or lilac-coloured blossoms, its narrow leaves, with their grey under-surface, which are said to thrive best by the seaside. Its botanical name is compounded of two Latin words signifying *sea dew*. In olden times the rosemary was esteemed an emblem of constancy. In *Hamlet* we read :

'There's rosemary for you, that's for remembrance ;'

And in the *Winter's Tale :*

'For you there's rosemary and rue ;
Grace and remembrance be with you.'

It was frequently used at funerals, and planted on tombs, and was deemed as essential in the wedding wreath as is now the orange blossom. It is said that Anne of Cleves wore it in her hair on the morning of her marriage with Henry. A rich scent or oil is distilled from the rosemary; and bunches of it used formerly to be used by our forefathers to stir their ale, and give a fragrant, spicy flavour to the beverage.

If the month is genial, the river side may be enlivened by a few early blossoms of the marsh-marigold (*Caltha palustris*), rearing its sturdy stems close by the water's edge.

> 'Oh brave marsh marybuds, rich and yellow!'

This brilliant golden flower, with its handsome round green leaf, oft decks the mountain streams of Scotland; and in Lapland it is greeted as the harbinger of spring, though it does not bloom there until May. It always chooses a moist, shady spot,

> ' Not the sunny plain,
> But where the grass is green with shady trees,
> And brooks stand ready for the kine to quaff.'

It is common in France, and flowers luxuriantly on the marshy lands of Holland.

And now we have culled nearly all February's flowers; and our bouquet for so wintry a month is not so small after all. If our homeward way lies through a clayey soil, we are likely to find an early blossom of the colts-foot (*Tussilago farfarus*), whose heart-shaped leaves do

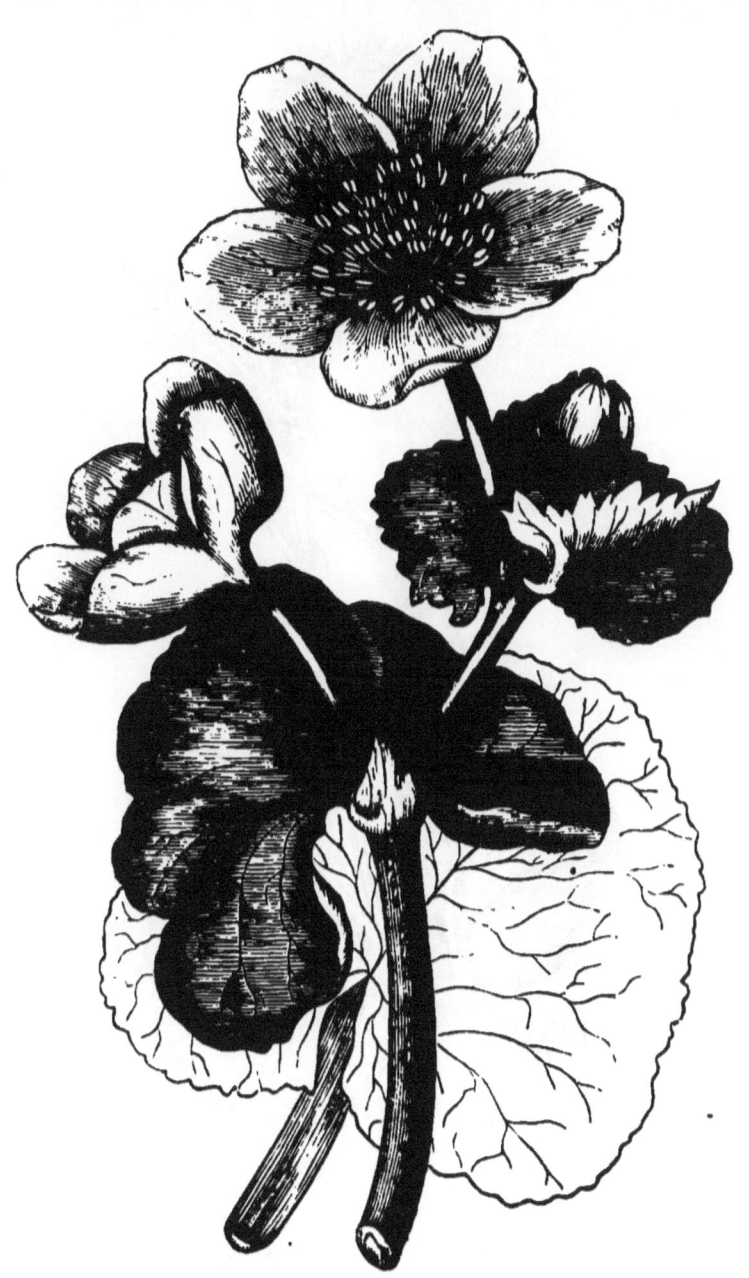

MARSH-MARIGOLD—*Caltha palustris.*

29

COLTSFOOT—*Tussilago farfarus.*

80

not unfold until the bright yellow flower has faded away.
It is well described by Bishop Mant:

> 'On scaly stem, with cottony down
> O'erlaid, its lemon-coloured crown,
> Which drooped unclosed, but now erect,
> The coltsfoot bright develops; decked
> (Ere yet the impurpled stalk displays
> Its dark-green leaves) with countless rays,
> Round countless tribes, alike in dye,
> Expanded.'

The cottony down, which is found on the under part of
the leaf, is gathered by country people for tinder; and
the feathery part of the seed is used by them for stuffing
mattresses. The leaves are often mixed in our British
tobacco, and the flowers infused as a remedy for coughs.
The botanical name is derived from *tussis*, a cough. I
am sure my young friends must all know the handsome
green leaf of the coltsfoot, so named from its fancied
resemblance to the foot or hoof of a colt, which may be
seen in every wayside walk; but I dare say few of them
ever thought how useful it could be made, though its
spreading roots and many seeds cause it to be a somewhat troublesome plant to many a farmer. Very beautiful are the ripe seeds of the coltsfoot, that float away on
their airy wings, and disperse themselves far and near:

> 'In nature's field take what you will,
> Each thing displays consummate skill.'

Much knowledge may be gained from the simplest

little flower, or humble tangled weed that creeps along the grass. How wonderful is its structure! Its perfect arrangement and adaptation to its position and requirements, from the bursting bud of spring, to the wafting of the downy seeds by the autumnal winds, all testifying to

'The well-ordained laws of Jehovah.'

'Whoso is wise and will observe these things, even they shall understand the loving-kindness of the Lord.' —Ps. cvii. 43.

III.

MARCH.

———◆———

'Strew, strew the glad and smiling ground
With every flower, yet not confound
The primrose drop—the spring's own spouse,
Bright day's-eyes, and the lips of cows.'

YOU often hear of the 'black winds of March,' and yet it is as often styled 'the merry month of spring.' Inconsistent as this seems, there are, indeed, days when bitter winds and sleety showers sweep across the meadows, checking and withering the bursting bud; and then, again, the warm, bright sunshine breaks forth, as though bidding nature rejoice that

'Winter has passed with its frowns away,
And the beautiful spring is coming!'

The merry-hearted bird, springing from bough to bough, twits his little song, or trims his feathered suit,

C

> ' Though tender leaves on tree and bush
>   Scarce hide the blackbird and the thrush.'

Let us, my young friends, choose such a sunny morning for our healthful ramble in the country. We shall find that vegetation has rapidly advanced since we took our last pleasant wander together, that nature is awakening from its winter sleep, and many fresh flowers await our gathering, in addition to the coltsfoot which now carpets the ground with its bright yellow flowers, the pale clusters of the spurge laurel, the gaudy dandelion flourishing everywhere, and the modest little spreading chickweeds. The violet, so prized in our last walk, may now be gathered in profusion, the air is perfumed with its fragrant sweetness; and the pure, innocent daisy, so loved by us all, is scattered thickly over the fresh green of the newly springing grass,

> ' As though some gentle angel,
>   Commissioned love to bear,
>   Had wandered o'er the greensward,
>   And left her footprints there.'

Truly no season is more lovely than our dawning spring, so worshipped by the poet, so welcomed by every lover of nature. Charles d'Orleans, who was taken prisoner at the battle of Agincourt in 1415, and remained for twenty-five years a poor solitary captive in England, wrote so delicately and beautifully of the return of spring, that I cannot resist giving a portion of his verse to my young friends :

'Now Time throws off his cloak again
Of ermined frost, and wind, and rain,
And clothes him in the embroidery
Of glittering sun and clear blue sky.
With beast and bird the forest rings,
Each in his jargon cries and sings ;
And Time throws off his cloak again
Of ermined frost, and wind, and rain.'

Wandering across the fields, we shall likely find many new pleasures greet us ; and perhaps the first in this month's list may be the lesser celandine, or pilewort (*Ranunculus ficaria*), whose golden blossoms,

'Shining like the glist'ning star,'

gleam brightly from hedge-bank or meadow-grass,

'The first gilt thing
That wears the trembling pearls of spring.'

Every lover of wild-flowers welcomes the celandine as one of the fairest and earliest of our spring blossoms.

'Ere a leaf is on the bush,
In the time before the thrush
Has a thought about its nest,
Thou wilt come with half a call,
Spreading out thy glossy breast
Like a careless prodigal ;
Telling tales about the sun,
When there's little warmth, or none.'

So sings Wordsworth, that true poet of flowers, whose verses have immortalized the celandine, which he claims as his own :

' There's a flower that shall be mine,
'Tis the little celandine.'

Beautiful as the celandine is, I grieve to confess it is a
very injurious plant to the land, and is said to destroy

LESSER CELANDINE—*Ranunculus ficaria.*

all others growing near it.   A large number of its blossoms spring from one root ; and its heart-shaped leaves

are spotted with a paler green. At night, and before rain, the flowers close, and open only in the brightest sunshine.

The major celandine (*Chelidonium majus*) is common in hedges, and blooms in May. There is also the sea celandine, or yellow-horned poppy (*Glaucium luteum*), which grows by the coast, and flowers in June.

But let us look for that 'earliest nursling of the spring,' that beautiful woodland flower, 'the primrose pale' (*Primula vulgaris*), so delicate in the sulphur tints of its pretty heart-shaped petals, set off to such advantage by its own handsome green leaf:

PRIMROSE—*Primula vulgaris.*

> ' On the wood's warm, sunny side,
>   Primrose blooms in all its pride.'

Who that loves spring sunshine and opening flowers,

does not welcome this fair blossom, which seems yet fairer for its early appearance in the very dawn of spring? Well does it merit its name of *prima rosa*, or *first rose;* for even in February, when we have ventured forth to pluck the delicate snowdrop, and ere we have rejoiced to see

> ' The vernal pilewort's globe unfold
> Its star-like disk of burnished gold,'

we may discover an early blossom ; nay, perchance, even

> ' A tuft of evening primroses,
> O'er which the mind may hover till it doses ;
> O'er which it well might take a pleasant sleep,
> But that 'tis ever startled by the leap
> Of buds into ripe flowers.'

Our English poets have paid due honour to this sweet spring flower, that unites such delicate beauty of colour, form, and fragrance. Milton speaks of the 'pale prim-rose :'

> ' The rathe primrose, that forsaken, dies.'

Ben Jonson calls it 'the spring's own spouse.' Words-worth speaks of the early passing away of this flower, which rarely continues to bloom later than May :

> ' Primroses, the spring may love them,
> Summer knows but little of them.'

There are several varieties, but the commonest is the one we have just gathered, and two lilac kinds, which latter are sometimes found in the north of England, and

Scotland, but do not bloom until July. Not one of the primrose tribe but what is rejected by all grazing animals, excepting swine. I have heard of primrose ointment being made from the flowers, and used as a cosmetic. The cowslip and oxlip both belong to this genus ; but of them we shall speak when they bloom.

We shall now find growing on the bank the little barren strawberry (*Potentilla fragariastrum*), so precisely similar in flower and leaf to the wood strawberry (*Fragaria vesca*), but differing, inasmuch as its pretty white blossom is followed by no fruit. It grows in patches, and loves to nestle beneath the shelter of a spreading tree. The wood strawberry is common throughout Great Britain, and grows in abundance in the woods of France. It does not blossom until May ; and I dare say you all know how sweet and refreshing is the fruit it bears :

> ‘ Fragrant, if small, and pleasant to the taste,
>   Agreeable in form and hue ;
>   With nought unwholesome, nought that seems misplaced,
>   Perfection in the strawberry we view.’

There is a delicate plant that sometimes puts forth its little green flowers as early as this, though more abundantly during next month. It is the tuberous moschatel (*Adoxa moschatellina*), an unobtrusive blossom, flourishing in moist woods and shady places.

> ‘ Adoxa loves the greenwood shade ;
>   There, waving through the verdant glade,
>   Her scented seed she strews.’

Its stems are solitary and upright, its flowers gathered in little yellow-green clusters, and emitting a musky

TUBEROUS MOSCHATEL—*Adoxa moschatellina.*

scent, which has led this plant by some to be called musk crowfoot. It flowers in March and April, ripening its berries in May, and scattering them as the poet has said.

Scarcely have the wintry winds died away ere the mezereon (*Daphne mezereon*) appears, one of the first blossoms to greet the returning spring, with its sweet odours, and the beauty of its delicate purple-pink wreaths. Its clustering flowers, which gather round the shoots of the former year, appear before the leaves are out, as Cowper says:

> ' Mezereon, too,
> Though leafless, well-attired, and thick beset
> With blushing leaves investing every spray.'

It is to be found in almost every part of Europe ; and, blooming at a season when flowers are few, it has become quite a garden plant. The branches make a beautiful yellow dye ; the acrid bark is valuable in medicine, and if bound down upon the skin, will raise a blister. Its root is often applied for toothache, but should be used cautiously, as it is apt to heat and inflame the mouth. Its scarlet, one-seeded berries, which in autumn cluster around the stems, are highly poisonous to man and beast, though harmless to birds ; and dearly do the thrush and blackbird relish them. This plant, and several of the Daphne, are often called laurel, from the similarity of their shining leaves to that shrub. It was formerly called *spurge olive*, and *mountain pepper ;* and the Italians have given it the endearing name of *biondella*—little fair one.

If we examine the earthy top of some old wall, we shall likely find, growing also in little patches, the small white blossoms of the common whitlow grass (*Draba verna*). Its flowers are cross-shaped, its stem only about two inches high, and a circle of delicate leaves surround its base. The whole plant is so small, that we must examine very carefully the low green moss in which it loves to dwell, ere we shall detect it.

Every child has noticed, throughout the summer, when wandering by the river-side, the large broad leaf of the butter-bur (*Tussilago petasites*), growing on a stalk some two feet high, and known to them under the name of 'umbrellas.' *Petasites* is taken from a Greek word signifying a broad covering, and is doubtless given to the butter-bur on account of its magnificent leaves, which are finely veined, and very hairy. It belongs to the coltsfoot tribe, and, like it, the flowers precede the leaves. In this month of March, the spike-like heads of downy, dusky, flesh-coloured blossoms appear, affording early food to the cheerful bee:

'While from their cells, still moist with morning dew,
The wand'ring wild-bee sips the honied glue,
In wider circle wakes the liquid hum,
And, far remote, the mingled murmurs come.'

How beautiful are the hedges now becoming with the white blossoms of the early sloe, or blackthorn (*Prunus spinosa*), which hang like snow-wreaths over its dark branches long before any leaves appear; but these take

the place of flowers as spring advances. What schoolboy does not relish in the autumn its harsh unwholesome
fruit, in flavour like an unripe damson, and which is
often used, I believe, in the adulteration of port wine!
Country people make a medicine from the infused leaves,
and say that the name of blackthorn is derived from the
fact of its blossoming during the 'black winds of March.'
Let us gather a branch for our nosegay, but take heed
of torn fingers, for its thorns are strong and sharp.

> 'Without a wry face, difficult it is
>   To eat the austere berries of the thorn :
>   Most difficult to penetrate the fence
>   With its sharp spines thick set on every side.'

I wonder if any of my young friends have ever observed
a flower familiarly known by the name of pickpocket,
but more correctly called shepherd's purse (*Capsella bursa-
pastoris*), on account of its little heart-shaped seed-vessels,
which are not unlike the form of a tiny purse, and which
are much more conspicuous than its clusters of small
white flowers. We must pluck a piece to examine as
we go along. It is to be found in almost any hedge or
field, and flowers nearly all the year.

And now, as we approach this old wall, how sweet is
the fragrance of the wallflower (*Cheiranthus cheiri*), that
greets us with its delicious odour! Its flowers are rich
in all the varied hues of red and yellow :

> 'The yellow wallflower, stained with iron-brown.'

We have only one British species of this shrubby plant, from whence proceed all our garden varieties. It is much prized in the East. It flowers early, heedless of March winds, and blooms throughout the summer, flourishing by the seaside cliffs of our southern coasts, or lingering near the old ruin. From this it has been called 'fidelity's flower:'

> ' When the proud tower and battlement
> By time are all decayed and rent,
> Then, in misfortune's gloomy hour,
> Springs up fidelity's sweet flower.'

The common dog's mercury (*Mercurialis perennis*) often puts forth its blossoms during this month, though, being green, we are apt to overlook them. It is a very common plant, speedily overrunning any neglected ground. Its flowers grow in spikes, on long slender stems; its leaves are handsome and abundant, forming thick patches of dark green. It grows about a foot high, blossoms the whole summer, and is exceedingly poisonous.

The graceful fritillary (*Fritillus meleagris*), which sometimes adorns the English meadow, but is by no means a common plant, blooms also at this season. It is a solitary pendent flower, its dark, tulip-shaped bell being curiously spotted with dull pink. There is just this one native of the family. Its name is derived from the Latin, *fritillus*, signifying a dice-box. In former times it was called Guinea-hen flower, on account of its spotted appearance; and a poet, in writing of it, has said :

> ‘ Chequered are thy leaves, as when
> Persecution's shadows fall
> On the paths of righteous men,
> Like a gloom funereal.’

The golden saxifrage (*Chrysosplenium oppositifolium*) may be sometimes gathered as early as this, growing by the water-side. It is a succulent plant, rising some few inches high. Its yellowish-green blossoms, nestling amongst its upper round green leaves, are small, and richly tipped with golden yellow.

The family of the saxifrage is numerous. Several species grow on rocks, stones, or dry walls, the commonest of which is the rue-leaved saxifrage (*Saxifraga tridactylites*), which has small white flowers, reddish leaves covered with sticky

GOLDEN SAXIFRAGE—*Chrysosplenium oppositifolium.*

hairs, and grows only a few inches high. The meadow saxifrage (*Saxifraga granulata*) is a handsome plant, growing a foot in height, with larger white flowers, and kidney-shaped leaves. Both these flower in May. But the best known of all is the species so common in the cottage garden, called none-so-pretty, and London pride (*Saxifraga umbrosa*). It is a native of Ireland, and indigenous also in England. Its flowers are a pinky white, dotted with dark red.

> 'The none-so-pretty is a lightsome flower,
> Called Amourette in France; and will not be
> Confined, but runneth even where it lists,
> And gives its heart to whom it liketh best.'

Blooming in the moist wood, or by the streamlet's edge,

> 'Drooping its beauty o'er the watery clearness,
> Wooing its own sad image into nearness,'

we may chance to find a first-born blossom or a clump of

> 'Daffodils,
> That come before the swallow dares, and take
> The winds of March with beauty,'

as Shakespeare beautifully expresses his allusion to the early flowering of this plant. He further tells us that,

> 'When daffodils begin to peer,
> Why, then comes in the sweet of the year.'

The single daffodil (*Narcissus pseudo-narcissus*), the daffadowndilly of former days, is by no means an uncommon flower:

> 'Beside the lake, beneath the trees,
> Fluttering and dancing in the breeze.'

Its drooping head is composed of a deep yellow cup, surrounded with a circlet of paler lemon-coloured petals. Its leaves are long, narrow, and a bright, deep green.

NARCISSUS—*Narcissus poeticus.*—DAFFODIL—*Narcissus pseudo-narcissus.*

It is rare, I believe, in Scotland, but in the south of England blooms freely in many a moist meadow, its usually brilliant golden blossoms sometimes assuming a more delicate cream-like tint, equally beautiful. It belongs to the narcissus tribe, named after Narcissus, a beautiful youth, who, as the poets tell us, falling in love with his own image in the water, pined away into a daffodil. But the poetical narcissus (*Narcissus poeticus*) is a fairer and more rare blossom than the daffodil. It has a snow-white flower, composed of six petals, with a yellow cup in the centre, edged with a fringe of deep purple or scarlet, and is scented with a strong, delicious fragrance.

> ' Narcissus, drooping on his rill,
> Keeps his odorous beauty still.'

It is a less hardy plant than the daffodil, and does not bloom until May, thus escaping the easterly winds that are so destructive to it.

> ' When the chilling east invades the spring,
> The delicate narcissus pines away.'

We have now, I think, gathered almost all our March flowers, and the waning sun bids us hasten homeward. As we go, let us observe how the hazel-tree is now decked with its hanging tassels, and the sombre alder is donning its dark, gloomy foliage. The flowers of the ash are coming out on its leafless boughs ; the spiry branches of the Lombardy poplar are unfolding their drapery ; and the beautiful green bursts forth from its winter shield in the well-cased buds of the horse-chestnut, that, ere summer dawns, will be

> ' Clad with blossoms white and fair,
> Blossoms that perfume the air ;
> Spreading wide, and towering high,
> Emblem of luxuriancy.'

How beautiful are the golden balls, or children's 'pussy-cats,' of the willow, which is usually called palm, and used in the religious ceremonies of Palm Sunday! In Ireland, where the church holds the anniversary of Christ's entrance into Jerusalem as a high festival, branches of yew are used instead of willow, and, after the priest's benediction, become 'blessed palms,' to be suspended in the church or home. Many are the associations of trees and flowers with the religious ceremonies of the church, as well as the celebration in olden times of great national sports and victories, when Beauty was crowned with myrtle, the patriot with oak,—when bays formed

'The victor's garland, and the poet's crown.'

And in later times each saint's day claimed its own particular flower: St. Valentine, the crocus; St. George, the harebell; St. Bartholomew, the sunflower; St. Patrick, the wood-sorrel; and we are told that

'The Michaelmas daisy, among the dead weeds,
Blooms for St. Michael's valorous deeds.'

Thus, when wandering through rural scenes, we are surrounded by the most interesting associations, awakened, it may be, by the humblest blossom that has sprung in our pathway, and which also, by the perfection of its beauty and structure, is ever telling of a supreme Creator, and proclaiming

'The hand that made it is divine.'

D

How often, also, when viewing the trees of the forest, or 'flowers of the field,' may we be reminded of portions of Scripture, and of Scripture truth! Our Saviour himself drew the attention of his disciples to the beauties of the earth *for this purpose:* 'Behold the grass of the field.' 'Consider the lilies.' Think of the mustard-seed, the barren fig-tree, the seed-corn and tares, and 'APPLY YOUR HEARTS UNTO WISDOM.'

IV.

APRIL.

———•———

April has come, the capricious in mien,
With her wreath of the rainbow, and sandals of green ;
Storms on her forehead, and flowers at her feet,
And many-toned voices, but all of them sweet ;
Playing, like childhood, with tear and with smile,
Weeping for ever, and laughing the while !
Months follow, fairer, when April is gone,
But none of the year have a gift like her own ;
Richer their colours, and sweeter their breath,
But no month of them all sees so little of death.'

OW lovely are our woodlands in this
showery, sunshiny month of April,
when the young-leaved boughs of
the forest trees, in their half-unfolded
beauty, are gently waving, and the
songs of the joyous birds ring merrily
through the woods !

' Up, let us to the fields away,
And breathe the fresh and balmy air ;
The bird is building in the tree,
The flower has opened to the bee,
And health, and love, and peace are there.'

Yes! spring has come at last,—delicate, fresh, emerald spring ;

> ' And earth exulting, from her wintry tomb
> Breaks forth with flowers.'

Many a lovely blossom is peeping from meadow and hedgeway ;

> ' Bells and flowerets of a thousand hues'

are spreading their gay carpet of variegated colours,

> ' As if the rainbows of the fresh green spring
> Had blossomed where they fell.'

The bluebell bends to every passing breeze ; the speed-well

> ' Gleams like amethyst in the dewy grass ;'

and the celandine—the beautiful celandine—glistens like gold from out the verdant meadow ; whilst the air is perfumed with the sweet odours of violets, ground-ivy, hyacinths, and many spring flowers.

> ' The welcome flowers are blossoming,
>   In joyous troops revealed ;
> They lift their dewy buds and bells,
>   In garden, mead, and field.
> From the green marge of lake and stream,
>   Fresh vale, and mountain sod,
> They look in gentle glory forth—
>   The pure, sweet flowers of God.'

There is not, however, a sweeter or more beautiful blossom in this month than the wood-anemone (*Anemone nemorosa*), that droops its modest head in such abundance

alike in wood and sheltered valley. It is the wind-
flower of our old poets,—'*l'herbe au vent,*' or wind-herb,
of the French. The English name is taken from the

WOOD-ANEMONE—*Anemone nemorosa.*

Greek word *anemos*, or *wind*, which was given by the
ancients, *why*, I am unable to tell you, unless it be that
ts delicate petals, so soon ruffled by the spring winds,

are often first found quivering in the fierce breezes of March. A poet has spoken of

> ' The coy anemone, that ne'er uncloses
> Her lips until they're blown on by the wind ;'

but I am more disposed to think it is to the warmth and brightness of the genial sunshine that the ' coy anemone' chooses to 'unclose her lips ;' for, when he pours his radiant beams with fullest light and vigour on the earth,

> ' Then, thickly strewn in woodland bowers,
> Anemones their stars unfold.'

What, my dear young friends, can be more exquisite than these fair, tinsel-looking, quivering blossoms, that nod and tremble amongst the grass, raising their slender stems, with graceful, drooping heads, amidst their triple circlet of dark, smooth, beautifully-cut leaves! The flower is white and star-shaped, delicately pencilled with purple lines ; the veins of the green leaves are tinged with crimson. And thus we see even this lowly blossom the divine Hand has streaked with loveliness and beauty.

Though continuing to bloom in rapid and plentiful succession, it is frail when gathered, and dies quickly ; thus we are told,

> ' Its beauty but awhile remains ;
> For those light-hanging leaves, infirmly placed,
> The winds, that blow on all things, quickly blast.'

There are various legends attached to the anemone.

Some poets have told us it owes its origin to Flora, the goddess of flowers and gardens, the privileged enjoyer of perpetual youth, who, jealous of the exquisite beauty of a Grecian nymph, changed her into this blossom. But the fable most commonly received is that connecting it with the death of Adonis, who was killed whilst hunting, and over whom Venus shed many tears, each tear that fell to the ground springing up a beautiful anemone. Now, my young friends may not *all* know that the ancient Greeks and Romans held a mythology, which is simply a collection of legends and fables, of course all fictitious, yet differing from fiction, because once believed to be an account of events that had actually taken place. Gods and goddesses were supposed to have presided over these events; and temples were raised to their memories, in which festivals were held, that often lasted for days and weeks. Now, Venus was honoured as the goddess of love and beauty, the mistress of the Graces, and queen of laughter. Adonis was her beloved favourite. He was passionately fond of hunting, and so bold and daring in the chase, that Venus, fearful lest in thus exposing himself he might one day be slain, forbade him to hunt wild beasts. But alas! Adonis, over-confident in himself, heeded not the injunction, and received a mortal injury from a wild boar. Then we are told the plentiful tears of Venus gave birth to the wood-anemone.

> ' The flower which sprang, as ancient fables tell,
> When 'neath the wild boar's tusk Adonis fell,

The youth beloved of Venus, from whose eyes
Poured crystal tears, like raindrops from the skies.'

The elegant, silky pasque-flower (*Anemone pulsatilla*) may also be gathered in this month, growing on banks or chalky pastures, but is not a common plant. Its delicate purple stars of flowers are larger than the wood-anemone, but do not grow so tall.

Another bright ornament of our spring meadows is that 'fragrant dweller of the lea,' the 'freckled cowslip' (*Primula veris*), of which the peasant-poet Clare has so beautifully written :

'Bowing adorers of the vale,
Ye cowslips delicately pale,
Upraise your loaded stems ;
Unfold your cups in splendour ; speak !
Who decked you with that ruddy streak,
And gilt your golden gems ?'

Exquisite indeed is this softly-fragrant, pale yellow blossom, with its handsome primrose leaf, and drooping flowers, or 'nectared bells,' supported on one delicate stem :

'Cowslips wan, that hang the pensive head.'

Hasten, my young friends, search far and wide for the largest, most perfect blossoms, and let us make a cowslip ball, by nipping off the clusters from the top of the stem, and passing them upon a string. Take care to press the flowers closely together, then draw them up into a ball, and tie the string securely. What a sweet-scented toy to toss into this lovely April air !

The cowslip, or 'lips of cows,' as Ben Jonson calls it, loves a moist soil and open situation. I have heard of wine being made from cowslips, and also of 'cowslip tea;' but I have little inclination to taste either. Montgomery, in his *Walk in Spring,* writes very beautifully of the cowslip and its uses, though he tells us nothing of what an injurious weed it becomes when abundant in pasture-land.

'Now, in my walk, with sweet surprise
I saw the first spring cowslip rise,
    The plant whose pensile flowers
Bend to the earth their beauteous eyes,
    In sunshine and in showers.

.   .   .   .

Lovely and innocent as they,
    O'er coppice lawns and dells,
In bands the rural children stray,
    To pluck thy nectared bells.

Whose simple sweets, with curious skill,
The frugal cottage dames distil,
    Nor envy France the vine;
While many a festal cup they fill
    With Britain's homely wine.'

The oxlip (*Primula elatior*) is also now in flower, but is somewhat rare. It greatly resembles the cowslip, but has larger flowers, spreading wider, and more upright, but not so tall. It has been called the great primrose, and the green leaves are much the same.

In this same moist meadow we are sure to find the

pale lilac, wan-hued cuckoo-flower (*Cardamine pratensis*).
Shakespeare's

> ' Ladies' smock, all silver white,
> That paint the meadow with delight ;'

thus named by the great dramatist from its fancied
resemblance, when blooming thickly, to linen spread
upon the grass. The flowers, when first they open, are
tinged with a delicate purple, but, when exposed to the
heat of the sun, gradually fade to a silvery white. Its
leaves are different at the root to those on the stem ;
but all are small, though the flower often springs a foot
high. Its foliage is pungent. It is called cuckoo-flower,
because it usually blooms when that bird begins to sing.
There are five species, but only one common as this,—
that is, the small, white-flowered, hairy, bitter cress (*Cardamine hirsuta*), the leaves of which are often gathered
for salad. Like the cuckoo-flower, it loves best the
moist meadow, or borders of a stream,

> ' Or where old winter leaves her plashy slough,
> The lady's smock will not disdain to grow.'

What a grace is now added to our hedgerows by the
dark blue flower of the lesser periwinkle (*Vinca minor*),
its large glossy leaves glistening in the sunshine, so
bright in their green as to give it the name of 'little
laurel !'

> ' The fresh periwinkle, rich in hue.'

Though not generally a common plant, it is most abun-

CUCKOO-FLOWER—*Cardamine pratensis.*

dant in some parts, and, spreading rapidly, soon covers large plots of green.

> ' Through primrose tufts in that sweet bower,
>   The periwinkle trailed its wreaths;
> And 'tis my faith that every flower
>   Enjoys the air it breathes,'

writes Wordsworth. Its name is derived from *vincio*, to bind, in consequence of its tough, trailing stems. Very lovely is this rich, many-flowered plant, which blooms all summer, its bay-like leaves flourishing far through the winter months.

> ' The lesser periwinkle's bloom,
>   Like carpet of Damascus' loom,
> Pranks with bright blue the tissue wove
>   Of verdant foliage.'

In Italy wreaths are made of the periwinkle, to deck the graves of infants. There is a white variety, also very beautiful, but not quite so plentiful as the blue kind. The larger periwinkle (*Vinca major*) does not bloom before May. It is much less common than the lesser periwinkle, and, as its name implies, both flowers and leaves are larger, the whole plant more erect.

Also, now trailing its long stems by every hedge-side is the fragrant ground-ivy (*Glechoma hederacea*), its blue or lilac flowers peeping out from the shelter of other plants.

> ' And there, upon the sod below,
>   Ground-ivy's purple blossoms show,
> Like helmet of Crusader knight,
>   In anther's cross-like form of white.'

Its blossoms grow in whorls of threes, and are supported at the base by a couple of dark, scallop-shaped leaves. In olden times the ground-ivy was infused for coughs,

GROUND-IVY—*Glechoma hederacea.*

and was regularly cried and sold in the streets of London. Both flowers and foliage make it a very pretty plant; but it is one most injurious to pasture-land.

The cranesbills are beginning to bloom in this month; and every one must recognise the little dove's-foot cranesbill (*Geranium molle*), which blossoms on almost every bank and waste place. It has long spreading stems, and round grey-green leaves, deeply cut into segments, and downy as velvet. Its flowers are small, and of a purplish red. There are seventeen species of cranesbill, but this blooms earliest of all. None, however, are more beautiful than the herb-Robert, or poor robin (*Geranium Robertianum*), which blooms next month,—a lovely flower, that shows its pretty pink blossoms in all places; and when these have died away, its leaves, in autumn often assuming a rich crimson glow, are as beautiful as the flowers themselves. Unfortunately it has a strong, unpleasant smell, and on that account is called *stinking* cranesbill. It is stated to have been called after Professor Roberts of Oxford, but many believe its name to bear a much older date. The name *geranium* is derived from the Greek, and signifies *a crane*. The seed-vessels of the plants, from being long and pointed, are supposed to resemble the bill of a crane.

Ah! here we have come upon the cuckoo-pint (*Arum maculatum*)—

'The hooded arum, early sprouting up'—

better known by its familiar names of lord-and-ladies

and wake-robin; in France fancifully called *bonnet de grand prêtre*, or high-priest's mitre. From the centre of its large, broad, glossy leaves, spotted with a darker green, rises a sort of column of lighter green. Carefully unrol this leafy sheath, and you will find a delicate, rich violet pillar, tapering at the summit, and surrounded by rows of white and red beading,—the whole being as elegant in form as it is beautiful in colour.

> ' Oh, cuckoo-pint, toll me the purple clapper
>   That hangs in your clear green bell.'

In winter a thick cluster of bright orange berries surround the stem, and are very conspicuous. Though highly poisonous, they are eaten, and seemingly relished by birds. The root of the plant is about the size of a nutmeg, and contains a farinaceous powder, not a bad substitute for flour. It is sold in Portland Island and Weymouth; and in London you can buy it under the name of Portland sago. The starch made from the root was highly prized in olden times, when ladies, and gentlemen too, were attired in ruffs and frills, so large and stiffly starched, that one wonders how they contrived to move either head or hands.

But we have not time to linger, for many April blossoms are yet to gather. There is that 'little western flower,'

> ' The shining pansy, trimm'd with golden lace.'

Shakespeare tells us that

> ' Maidens call it love-in-idleness.'

This pretty little wild heartsease (*Viola tricolor*), so common during the summer months, is a species of violet, its name of pansy being a corruption of the

HEARTSEASE—*Viola tricolor.*

French *pensèe*—thought. 'There's rosemary,' says poor Ophelia, 'that's for remembrance ;—pray you, love, remember ; and there is pansies, that's for thoughts.' It is rich in names given by the old poets, such as, Jump-up-and-kiss-me, Herb-trinity, Pink-of-my-john, Three-faces-under-a-hood, etc. It is a general favourite: Spencer speaks of 'the pretty paunce;' Dryden of 'pansies to please the sight;' Hunt mentions the

> ' Heartsease, like a gallant bold,
> In his cloth of purple and gold.'

But Milton gives the finest description:

> ' The pansy *freaked* with jet.'

It blossoms on banks and cultivated fields, and varies much in colour, being sometimes pure white, at others a delicate cream, or tinged with yellow, blue, and purple.

Wandering through the woodlands, we cannot fail to notice a small white, delicate, bell-shaped flower, which blooms freely in the shady place, yet may often be found decking the high mountain. It is the pretty wood-sorrel (*Oxalis acetosella*), which never appears earlier than April, though frequently growing in cold countries. It was found by Captain Parry in places where scarcely any other flower ventured to blossom; and we are told that, 'when he saw no longer the wood-sorrel, he had reached the dreary regions of perpetual snow.' It is a humble little flower, lowly in growth, its delicate pearl-white petals elegantly veined with purple lines:

> ' The trim oxalis, with her pencilled leaf.'

Almost as beautiful is its bright green triplet leaf, shaped like three small hearts joined together at the points, and

E

which spring profusely around the blossoms. It is the most sensitive wildling we have ; for so soon as the evening dews begin to fall, it droops its leaves around the stems, and ever seems to shrink at the approach of night, or the faintest whisper of a coming storm. Its

WOOD-SORREL—*Oxalis acetosella.*

roots resemble coral beads strung together, as Charlotte Smith describes it :

> ' The wood-sorrel, with its light green leaves,
> Heart-shaped and triply-folded, and its root
> Creeping like beaded coral.'

Its seeds are each cased up in a sort of elastic covering, which, when fully ripe, suddenly bursts open, and ejects its contents to a great distance. Its leaves contain a most agreeable acid, resembling lemon, and in consequence are commonly eaten on the Continent in salads and fish-sauce. The salts of lemon, which is so often used to remove iron or ink stains from linen, was formerly made from a preparation of the juice extracted from its leaves.

There were many old names to the wood-sorrel; amongst others, it was formerly called cuckoo's meat, because it was thought the cuckoo loved to feed upon it; and Alleluya, from the veneration in which the plant was held. A peculiar reverence was ever felt for the number three; and therefore the fact of its being a trefoil was sufficient to proclaim it holy. Even as early as the time of the Druids, its triple leaf was looked upon as an emblem of the mysterious Three in One, which they ever endeavoured to illustrate in their worship; and their reverence was increased by the crescent mark on each fully matured leaflet, in which they beheld an emblem of the moon, another of their symbols. It is the *true* shamrock of Ireland, and the undoubted plant from which, as was said, St. Patrick, the first missionary sent to that country, plucked one of its pretty trefoil leaves, when illustrating the doctrine of the Trinity to his disciples.

The trefoil, which many erroneously suppose to be

the original shamrog, adopted by the Irish as their national emblem, and the blossom which is now worn by them on St. Patrick's day, their patron saint, is the white Dutch clover (*Trifolium repens*), which, as well as the sweet-scented, common purple clover (*Trifolium pratense*), is so valuable an herbage plant to the farmer:

> ' Sow in good time the trefoil, that in spring
> Will juicy herbage to the cattle bring ;
> Cut it, and dry it for the winter rack ;
> Be provident, and thou shalt nothing lack.'

They are called trefoils on account of their triple leaflets. Observe how the bees swarm and hum over the clover field,

> ' Flying solicitous from flower to flower,'

and sucking the honey from each full blossom. There are seventeen species of wild trefoils,—purple, yellow, and white. One, very common (*Trifolium filiforme*), has small yellow flowers, which bloom on almost every country wayside. There are many old superstitions connected with the clover, which was supposed to possess a peculiar charm against witchcraft. Many a peasant, in past days, wore a spray to keep away the witches, and preserve him against all magic arts ; and even yet there seems a lingering belief in its virtue, for we are all apt to exclaim at our luck when we chance to find a four-leaved clover !

There is not a more lovely flower to be gathered during this month than the beautiful blue-bell, or wild hyacinth (*Hyacinthus non-scriptus*), which blossoms so abundantly in our woodlands, and continues, as Keats has said,

> 'The shaded hyacinth, alway
> Sapphire queen of the mid-May.'

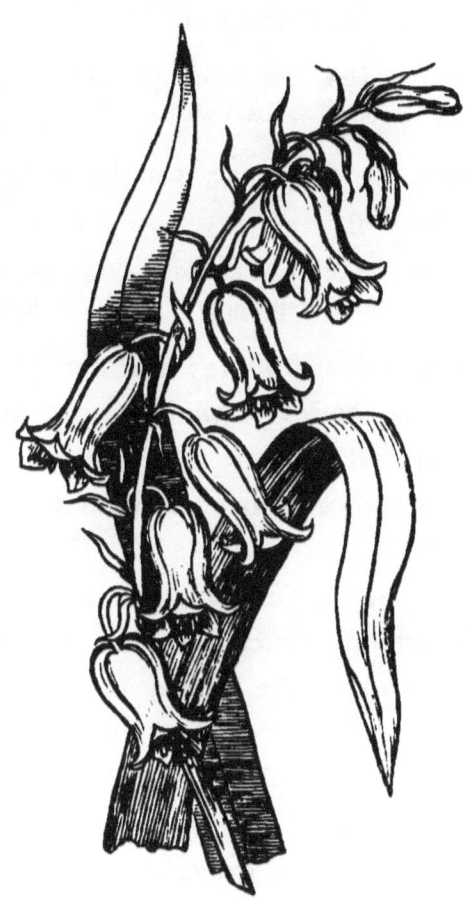

HYACINTH—*Hyacinthus non-scriptus.*

How modestly does each spike of flowers droop from the thick, succulent stem, that seems to bend beneath its

weight of beauty, and seeks to hide amidst its abundant
sword-like leaves

'The languid hyacinth,'

as one of our poets has called it. How pure is the deep
blue colouring of its blossoms!

'Hyacinth, with sapphire bell
Curling backwards.'

How delicate its fragrance, caught up by the sweet
spring air, scenting

'The lone copse or shadowy dell,
Where clustered knots of blue-bells blow!'

Sometimes it is found white; and our garden varieties
vary in all the shades of pink, purple, and blue; but
these roots are all brought from eastern countries, and
are much more powerful in scent than our common wild
hyacinth

The whole plant is full of a poisonous clammy juice,
from which formerly starch and glue were made. Poets
have loved to extol the beauty of the hyacinth, that
blooms in the midst of delicate anemones, 'pale prim-
roses,' and fragrant violets. Most beautifully Shakespeare speaks of it as one of the 'fairest flowers' with
which to sweeten the sad grave of poor Fidele:

'Thou shalt not lack
The flower that's like thy face, pale primrose, nor
The azured blue-bell, like thy veins; no, nor
The leaf of eglantine, whom not to slander,
Outsweeten'd not thy breath.'

A legend is told of how this flower was named after Hyacinthus, a youth whose education had been entrusted to Apollo, by whom he was greatly beloved. Whilst playing together a game of quoits, Zephyrus, one of the winds, who was jealous of the affection existing between them, blew the quoit of Apollo on to the head of Hyacinthus, who was instantly killed with the blow. Apollo, disconsolate at the death of his favourite, changed his blood into a flower, which has ever since borne his name:

> ‘Apollo with unweeting hand,
> Whilome did slay his dearly loved mate,
> Young Hyacinth, the pride of Spartan land ;
> And then transformed him to a purple flower.’

The ancient poets also inform us that the leaves of the plant were streaked with black, emblematic of sorrow ; and Milton has spoken of it as

> ‘That sanguine flower inscribed with woe.’

but we look in vain for such lines in our beautiful wild hyacinth, and consequently it has been named *Hyacinthus non-scriptus*, which signifies, *not written.*

One can scarcely wander near a corn-field without finding a spray of fumitory (*Fumaria officinalis*). Its name is derived from the Latin *fumus* (smoke); but *why*, I am unable to tell you. Its smell, though unpleasant, is in no way similar to smoke. Its small, deep, purple, rose-tinged flower, has a black spot on it, is tubular in form, and drooping, grows in a spike-like fashion up the stem.   Its leaves are branched, or divided into segments, and are of a beautiful blue-green colour.

That it is somewhat of a pest to the husbandman, we may infer from the rhyme, which perhaps also explains the significance of its botanical name :

> ‘ Smoke of the earth we fumitory call,
>   For like a vapour it o’erspreadeth all
>   The ripening corn-field ; and the farmer’s spleen
>   Is roused to see it creep the stalks between.’

But if hated by the farmer, it was not so in former days by the country maid, who valued it as a cosmetic, though holding it in some awe, on account of many superstitious associations connected with this plant :

> ‘ Fumitory too, a name
>   Which superstition holds to fame ;
>   Whose red and purple mottled flowers
>   Are cropped by maids in weeding hours,
>   To boil in water, milk, or whey,
>   For washes on a holiday,
>   To make their beauty fair and sleek,
>   And scare the tan from summer’s cheek.’

The lovely blue flowers of the Germander speedwell (*Veronica chamædrys*) are now sprinkling the grassy banks with their short-lived abundant blossoms, or gleaming brightly from every wayside, as though bidding ‘ good speed’ to every passer-by. It has been called *bird’s-eye*, from the white rays that mark the centre of its flowers, which grow spike-like up the stalk :

' The little veronica called the bird's-eye,
  It hath a blue tint like a summer sky :
  True image the Greeks termed it ; why should not we
  In its beauty and grace fair resemblance see ?'

The rose-like green leaves are notched, and grow in pairs up the stem.

A commoner species is the ivy-leaved speedwell (*Veronica hederifolia*), which has small, delicate, paler-blue blossoms, and thick ivy-shaped leaves. It has long, entangling stems, which spread rapidly over the corn-fields, creeping so closely to the ground as often to escape observation.

But surely we have gathered a sufficient store of flowers to-day; both limbs and minds will be growing weary. It is time to bend our steps homeward ; yet, before saying good-bye to the woods, let us raise our heads and take a glance at the many stately trees that blossom in this month—the oak, chestnut, ash, willows, and birch ; the wild pear, box-tree, etc.; and look at that bright mass of foliage that twines and festoons itself from branch to branch, and down amongst the low shrubs and hedges. It is the red-berried bryony (*Bryonia dioica*), one of our most graceful climbers, that weaves its pretty white blossoms veined with green, its rich vine-like leaves, and curling tendrils, through every thicket :

' The bryony,
So lavish of its vine-like growth.'

Well does it merit its old name of white vine, from the

luxuriant richness of its foliage. In autumn it is covered with hairs, and adorned with deep red berries, which are said to be poisonous, but are relished by birds. Its white root is full of starch, which, when dried, yields a flour that has been found useful to the poor in times of scarcity. Do my young friends know that there have been seasons of famine, when the poor peasantry have been compelled to subsist on wild roots and seeds, sorrel, boiled nettles, etc.? Job speaks of mallows and juniper-roots being eaten at such times:

> ' For want and famine they were solitary.'
> 'Who cut up mallows by the bushes, and juniper roots
>   for their meat.'
>
> Job xxx. 3, 4.

Ought not our hearts to be full of gratitude that God has provided so amply for *us;* that in our days such times of dearth are unknown; that He has made us rich in plants for food and· clothing, shelter and shade; gladdened our lives by the fair verdure of the fields, and the fairer flowers that adorn them? For

> ' He might have made the earth bring forth
>   Enough for great and small,
> Enough of corn, and wine, and oil,
>   Without a flower at all.
>
> He might have made enough, enough
>   For every want of ours,
> For medicine, luxury, and for toil,
>   And yet have made no flowers.'

Let us, then, thank Him also for the profusion of summer blossoms, which scatter their buds of beauty on every side, as fresh and fair as when Adam first beheld them in Eden.

V.

MAY.

'Hail! beauteous May, that dost inspire
Mirth, and youth, and warm desire ;
Woods and groves are of thy dressing,
Hill and dale doth boast thy blessing !'

ANOTHER month has passed away since we enjoyed our last sunny ramble together, my dear young friends; and

'Flowery May, who from her green lap throws
The yellow cowslip and the pale primrose,'

has arrived, with a host of fresh blossoms to gather. Hasten, and let us away to the beautiful woodlands, so joyous with many voices hailing the advent of summer ; for

'May is the very month of mirth !
And if there be a time on earth
When things below in part may vie
For beauty with the things on high,

77

> 'Tis in that balmy vernal time,
> When nature revels in her prime,
> And all is fresh, and fair, and gay,
> Resplendent with the smiles of May.'

Opening flowers and bursting buds greet us on all sides,

> ' As if the dew of April showers
> Had woo'd the sun, and turned to flowers.'

The trees are so rich in their emerald green, that the light is scarce seen through their quivering leaves. All nature is fresh, and fair, and beautiful, filling the heart with gladness :

> ' The little brooks run on in light,
>   As if they had a chase of mirth ;
> The skies are blue, the air is warm,
> Our very hearts have caught the charm
>   That sheds a beauty over earth.'

The mountain-ash, the service-tree, the wild-pear, and crab-apple, which all belong to the *Pyrus* family, are now in bloom. Very beautiful is the mountain-ash (*Pyrus aucuparia*), whether we see it in spring, adorned with its bunches of white flowers and graceful feathery leaves, or in autumn, when decked with glittering clusters of rich coral berries, so loved by birds.

Strong indeed is the honey-like fragrance emitted by the service-tree (*Pyrus torminalis*), and very handsome are its white hawthorn-like blossoms, and large dusty leaves, in shape not unlike the sycamore, but looking as though they had grown by the door of a corn-mill.

The wild-pear (*Pyrus communis*) is a tall tree, with thorny branches and snow-white blossoms. The crab-apple (*Pyrus malus*) is a small tree, with spreading branches and white blossoms exquisitely tipped with rose-colour.

These two last are the origin of all our cultivated varieties.

Many of the *Prunus* family are also now in flower, of which the sloe is one, which we admired in March. The cultivated plum, cherry, damson, etc., have all sprung from this genus. Very pretty are the white blossoms of the bird-cherry (*Prunus padus*), springing from their fresh green leaves. And equally beautiful is the wild-cherry (*Prunus cerasus*) :

> ' Oh, there never was yet so fair a thing,
>   By racing river or bubbling spring,
>   Nothing that ever so gaily grew,
>   Up from the ground, when the skies were blue ;
>   Nothing so brave, nothing so free,
>   As thou, my wild, wild cherry-tree.'

The hawthorn-tree (*Cratægus oxyacantha*) is now lovely in the beautiful bridal attire of the sweet perfumed haw-thorn or May :

> ' Famed 'mid vernal scene
>   For gracing May's propitious hour
>   With prodigality of flower,
>   Pink-anthered 'mid its petals pale,
>   And lending fragrance to the gale,
>   Hailed from its fair and sweet array,
>   The namesake of the lovely May.'

Few trees can exceed the hawthorn in beauty, or afford so cool a shade in sunny fields. Who knows it not? Who loves it not?

> ' My childhood's earliest thoughts are linked with thee :
>   The sight of thee calls back the robin's song,
>   Who, from the dark old tree
>   Beside the door, sang clearly all day long ;
>   And I, secure in chishish piety,
>   Listen'd as if I heard an angel sing
>   With news from heaven, which he did bring
>   Fresh every day to my untainted years,
>   When birds, and flowers, and I were happy peers.'

The name *hawthorn* is said to be a corruption of the Anglo-Saxon *haegthorn*. Its blossoms vary from a delicate white to a beautiful deep pink or blush colour ; and the difference in tinge is the effect of the soil in which it grows. Its haws afford an abundant supply of food to the glad songsters of the woods. There used to be an old custom of 'going a-Maying' on May morning, when the villagers went to the woods, and cut branches of flowering hawthorn to suspend over their doors. This was supposed to be the remnants of a heathen rite, which was paid to Flora, the goddess of flowers. The hawthorn-tree lives to a great age. I have seen one in a garden near Edinburgh, which is said to be upwards of six hundred years old. It reminds one of Wordsworth's verse :

> ' There is a thorn, it looks so old,
>   In truth, you'd find it hard to say

How it could ever have been young,
It looks so old and grey.'

Creeping and winding through the intricacies of the hedge, we observe the greenish-white clusters of wild clematis (*Clematis vitalba*), sometimes called traveller's joy, and virgin's bower, its luxuriant stems often extending twenty feet :

'The traveller's joy is a darling thing ;
None loveth it more than I.
I've seen it in courtly gardens cling,
I've seen it 'mid rocks and ruins spring,
I know hedgerows where 'tis wandering,
And I smile as I pass it by.'

Though it adds beauty to the trees and bushes it entwines, it is a very destructive plant, and often injures, and even strangles them. It is, however, one of the most graceful climbers we have, twining and decorating our hedges, which are rendered still more beautiful by the fragrant waving blossoms of the honeysuckle or woodbine (*Caprifolium periclymenum*) :

'So doth the woodbine, the sweet honeysuckle,
Gently entwist ;'

mingling its sweet breath, and mixing its streaky yellow and red blossoms, with every bush :

'Copious of flowers, the woodbine, pale and wan,
But well compensating her sickly looks
With never-cloying odours, early and late.'

F

After the blossoms have faded away, clusters of deep red berries appear to deck the autumn hedge, and afford food to birds. There are two other species of woodbine,

HONEYSUCKLE or WOODBINE—*Caprifolium periclymenun:.*

but neither so common as this. All our poets have loved the honeysuckle. Chaucer speaks of it as 'the fresh woodbine ;' Milton, 'the flaunting honeysuckle ;' Barry Cornwall, 'the sweet-breathed, frail-perfuming woodbine ;' but Leigh Hunt, in one half-line from Rimini, *describes* it :

> ' The suckle's streaky light.'

Ben Jonson tells us how

> ' The honeysuckle doth entwine
> Itself with bryony and jessamine.'

And here we find this graceful, elegant climber, the black bryony (*Tamus communis*), twining its long stems round the branches of bush and tree, weaving its wreath of bright leaves, even to the highest bough :

> ' The slender bryony, that weaves
> His pale green flowers and glossy leaves
> Aloft in smooth and lithe festoons.'

Its flowers are insignificant ; its poisonous berries, green during summer, gradually ripen and redden as autumn approaches. It gains its name of *black* bryony from its root, which, however, is white inside, and full of a starchy substance.

Down by the hedgerow we shall find the greater stitch-wort (*Stellaria holostea*), sometimes called white-flowered grass, also stellaria or star-flower, from its form. It is a tender, brittle plant, with leaves of a delicate green, the flowers pure white, and rather less in size than a prim-rose. There are several kinds, but none commoner than this, which decks every wayside.

Near it we may discover an early blossom of the common enchanter's nightshade (*Circœa lutetiana*), a pretty plant, with small pinkish flowers and heart-shaped leaves, loving best a home in a shady place.

Here, also, in the hedgeside, we may find the columbine (*Aquilegia vulgaris*), which is usually of a blue tinge, but which I have often found both red, white, and of a pink fleshy colour. W. Browne speaks of it as

'The white, the blewe, the flesh-like columbine.'

None of its tints, however, are either rich or brilliant. L. A. Twamley says:

'The columbine? Full many a flower
  Hath lines more clear and bright,
Although she doth in purple go,
  In crimson, pink, and white.'

COLUMBINE—*Aquilegia vulgaris.*

It is a delicate pendent blossom, with five petals, elongated at their base into a horn-shaped spur. These have been compared to the drops and bells of the fool's cap; consequently the flower has been made the emblem of folly:

> ' Folly's cap and bells in thee,
> Columbine so fancy free,
> Foolish folk declare they see.'

Shakespeare, in *Hamlet*, tells us that

> ' Columbines mark ingratitude.'

And we are elsewhere told it is the emblem of hope to the deserted:

> ' The columbine, in tawny often taken,
> Is then ascribed to such as are forsaken:
> Flora's choice buttons of a russet dye,
> Is hope even in the depth of misery.'

It was formerly much used in the decking of churches at Whitsuntide; and, in consequence, the flower is still considered emblematic of that season. Its leaves are a deep green, notched and cut into segments, the stems dark, and often very rich in colour.

The major celandine (*Chelidonium majus*) blooms in this month. Its botanical name is derived from the Greek signifying *swallow;* and for this reason it was formerly called swallow-wort. It is quite unlike the celandine of Wordsworth, being much taller, having large-lobed leaves, and its seed-pod protruding conspicuously from the centre of its bright yellow flower.

We may also now gather the pretty pink feathery flowers of the ragged robin (*Lychnis floscuculi*),

'With its piped stem streaked with jet;'

the handsome blossom of the red campion (*Lychnis dioica*), which grows some two feet high, and often absolutely illuminates our hedges with its bright-coloured flowers; the yellow weasel-snout (*Galeobdolon luteum*), or archangel, as it is often called,—a tall, slender plant, with bright yellow blossoms mottled with red, and growing in whorls; the common mouse-ear hawkweed (*Hieracium pilosella*), which you are so likely to mistake for a dandelion, until you examine its bright lemon blossoms, occasionally striped with red, and its hairy leaves with their grey under-surface; and the little insignificant field-madder (*Sherardia arvensis*), that hides itself in the grass, with its small leaves growing in star-like circles, and its modest tufts of lilac flowers.

We must not, however, linger by the wayside, and neglect the host of blossoms that await us elsewhere, reluctant as we may be to quit the fields and hedges, that are now just bursting into all their summer richness; where

'The waving trees
Throw their soft shadows on the sunny fields,
Where, in the music-breathing hedge, the thorn,
And pearly-white May blossom full of sweets,
Hang out the virgin flag of spring, entwined
With dripping honeysuckles, whose sweet breath

RED CAMPION—*Lychnis dioica.*

87

> Sinks to the heart, recalling with a sigh
> Dim recollected feelings of the days
> Of youth and early love.'

We will hasten down to the streamlet's side, and gather the tall handsome blossoms of the yellow iris (*Iris pseud-acorus*), without which our May bouquet would be quite incomplete. But beware of boggy ground; for wet feet may bring both a chiding and a cold. The iris, or corn-flag, as it is often called, is one of the rivulet's most beautiful ornaments, with its bright yellow flower, so exquisitely pencilled with deep brown, and its mass of long, dark-green, sword-like leaves:

> 'Amid its waving swords, in flaming gold
> The iris towers.'

I know of but one other British species,—it is the stinking iris (*Iris fœtidissima*), which is also a handsome plant of a dull purple colour, but, as its name implies, possessing an unpleasant odour, yet is often called roast-beef plant! Its scarlet berries, bursting from the seed-pod, are very bright and beautiful during the winter months.

> 'Thou
> Makest no concealment of thy treasure-seeds,
> But show'st them openly to every eye.'

The iris is by some said to be the *fleur-de-luce* chosen by Louis VII., king of France, as his heraldic emblem, and called *fleur-de-Louis*, from which has sprung our corruption of *fleur-de-lis*.

The summer snowflake´ (*Leucojum æstivum*) blossoms in moist ground during this month; and very exquisite are its pure blossoms, bursting in clusters from their dried, upright, solitary spatha or sheath, like ready-culled bunches of snowdrops.

And now let us wander across the meadow, where the gay golden buttercups (*Ranunculus bulbosus*) gleam in profusion,— Shakespeare's

'Cuckoo buds of yellow hue;'

and under this clump of shadowing trees 'consider the lilies of the field' (*Convallaria majalis*),

'Our England's lily of the May,
    Our lily of the vale.'

See how prettily the sweet, pure, modest spike of bells peeps from amongst its handsome, broad green leaves,

'The lady lily, looking gently
    down!'

How graceful is its form, how sweet its scent, how exquisite its purity!

BUTTERCUP—*Ranunculus bulbosus.*

'The lily, of all children of the spring,
The palest, fairest too, where fair ones are.'

And, behold, also growing
by the root of this tree,

'Is the sweet little woodruffe; how
  lowly it grows
Beneath the rough bramble and
  tangled wild rose !
The sly little beauty ! half hidden,
  half seen,
Gleam her fairy-like flowers, and
  ruffs of pale green.'

The sweet woodruff (*Asperula
odorata*) has small white clustering flowers, and fresh green
leaves, growing in rings or
whorls around the stem.

' The woodruff lifts her fragrant crown
  Of star-like blossoms, pure as snow,
  With radiant fringe of leaves below,
    In greenwood shade.'

The dried leaves of these flowers,
on account of their scent, are
often mixed, I believe, with
snuff.

Common now in the wood is
the green tway blade (*Listera
ovata*), an orchis, with spike of

Sweet Woodruff—*Asperula odorata.*

small, yellowish-green flowers, and strong, broad, ovate

leaves, growing in couplets. It belongs to a numerous family; but, as few bloom before the end of May, we will not gather any until next month.

During this month the common bugle (*Ajuga reptans*) blooms in moist parts of the wood, so serviceable in healing a cut finger. It is a hardy plant. Its purple blossoms, varying to a lilac or white, mix with its short, broad leaves, growing up an erect stem. There are four species of this plant. One we may perhaps find—the ground pine, or yellow bugle (*Ajuga chamæpitys*)—as it is common, and blooms in this month. Its flowers are yellow, spotted with red, its leaves long and narrow.

There, on this grassy bank,

> ‘ The strawberry weaves
> Its cornets of threefold leaves
> In mazes through the sloping wood.’

This is the common wood-strawberry (*Fragaria vesca*),— a sweet, rose-like blossom, its delicate white petals timidly peeping from amongst its pretty green leaves. Unlike the strawberry (*Potentilla fragariastrum*) which we gathered in March, its blossom is followed by a sweet and wholesome fruit, which grows plentifully in the woods of France, where it is by many esteemed superior to the cultivated strawberry.

There, also, we find the herb Paris (*Paris quadrifolia*), more singular than beautiful, one of our largest *green* flowering plants. It is called also the love-knot, from the curious way in which four large leaves gather in the form of a cross around the stem, from the summit of which rises its four-leaved flower, in a cup formed also of four leaflets. Like nearly all green flowers, it is of a poisonous nature.

And now, what is mantling this old wall, springing from each crevice, throwing its thread-like branches, its small, ivy-shaped, rich green leaves, and many lilac blossoms over the stonework, as though to cover or deck its decay? It is the ivy - leaved toadflax (*Linaria cymbalaria*), familiarly called mother - of - thousands, from the profusion of its blossoms. The tiny flower is shaped like a snapdragon, and of a pale lilac colour, marked with yellow. The leaves are tinged with reddish purple on their under surface. There are six species.

Beside it we shall likely find the common pellitory of the wall (*Parietaria officinalis*), its

IVY-LEAVED TOADFLAX—*Linaria cymbalaria.*

little reddish-purple blossoms crowded closely between
the stem and leaves. It is often to be found near the
sea.

EUROPEAN CHICKWEED—*Trientalis Europæa.*

About this time the pretty European chickweed, winter
green (*Trientalis Europæa*), begins to show its elegant

little blossoms, which are of a brilliant white, and droop on their slender stems at the approach of night or rainy weather. Its leaves are a delicate green, the whole plant fragile-looking, as it bends and quivers with every toying breeze. There is but one British species.

Away on the moors we might find, now opening into beauty, the elegant little *Linnæa borealis*, that droops its fragrant bells with such grace and beauty, and is named after Linnæus, the great botanist.

The sweet gale, or Dutch myrtle (*myrica gale*), blooms during this month, on heath ground. It is a small shrubby plant, with a powerful aromatic odour not unlike the myrtle, which it also resembles in appearance, though a more crowded shrub, with lighter green leaves.

During May many vetches come into flower. They have delicate leaves, are nearly all twining plants, creeping amongst the branches of shrubs, and holding fast with their winding stems and clasping tendrils.

The pea-shaped blossoms of the tuberous bitter vetch (*Orobus tuberosus*) are of a blue or pink purple, marked with darker veins. The flowers have long stalks, the leaves growing in pairs on the stem. In Holland the roots are roasted as chestnuts, which they greatly resemble in flavour.

The grass vetch (*Lathyrus nissolia*) is a crimson flower, with slender grass-like leaves, destitute of tendrils.

The common or spring vetch (*Vicia sativa*) is found in pastures, and grows in the borders of fields.

The handsome yellow clusters of the common kidney vetch (*Anthyllis vulneraria*) are common on hills or sea-side cliffs. Its blossoms crowd together, two on each stem, and lie in a nest, as it were, of white silky wool.

The sweet-milk vetch (*Astragalus glyciphyllus*), sometimes called the false acacia, which is found in thickets, has a dull yellow, butterfly-shaped flower, with leaves larger than the other vetches.

There are many other varieties; but we have not time to linger. All are pretty, graceful plants. Their pods, so full of seeds, yield in abundance food for birds, and often for man also. It was from a species of vetch—the lentil (*Ervum lens*)—still prized and esteemed in Eastern lands, that the mess of red pottage was formed, for which Esau, weary with hunting, sinned, and sold his birthright. Several of the vetches are cultivated in our fields for herbage for cattle.

But what is that pure white cluster of blossoms which rises yonder amongst the grass? It is the beautiful ox-eye (*Chrysanthemum leucanthemum*), one of our commonest ornaments of the meadow. This tall, graceful flower, with its golden eye, is sometimes called the moon-daisy:

' Lo, the staring ox-eyes, plentiful are they,
 Gleaming in the pasture, where the children play ;
 Plucked up, and down trodden, scattered far and near,
 Spite of every obstacle, they spring up year by year.'

We may also now gather the pretty wood loosestrife or yellow pimpernel (*Lysimachia nemorum*), the

OX-EYE—*Chrysanthemum leucanthemum.*

' Yellow Lysimachus, to give sweet rest
  To the faint shepherd.'

Its delicate stems trail over the ground ; its leaves are a
bright green, its flowers a pure yellow ; and, altogether, it
is a sweet, attractive little flower.

But, my young friends, time fails me to introduce you
to all the floral acquaintances of this month—

' Bright gems of earth, in which perchance we see
  What Eden was, what Paradise may be.'

We will take a glance at the trees and shrubs, and then
it must be goodbye.   The beech (*Fagus sylvatica*), and
the Scotch fir (*Pinus sylvestris*), are both in flower now ;
the holly (*Ilex aquifolium*), of which we gossiped in
January, is in full beauty ; the air is becoming perfumed
with the powerful, luscious scent of the elder (*Sambucus
nigra*), just bursting into flower,—a somewhat clumsy
ungraceful tree, yet beautiful in its rich, creamy-white
blossoms ; the spindle tree (*Euonymus Europæus*) is
decked with its small greenish-white blossoms ; and
the white clusters of the dog-wood (*Cornus sanguinea*),
which bloom often throughout the entire summer, are
bursting out on the deep red, white-spotted branches of
the bush.

' Behold ! the trees now deck their withered boughs :
    Their ample leaves, the hospitable plane,
  The taper elm, and lofty ash disclose ;
    The blooming hawthorn variegates the scene.

The lily of the vale, of flowers the queen,
  Puts on the robe she neither sew'd nor spun ;
The birds on ground, or on the branches green,
  Hop to and fro, and glitter in the sun.

Now is the time for those who wisdom love,
  Who love to walk in virtue's flowery road,
Along the lovely paths of spring to rove,
  And follow nature up to nature's God.'

VI.

JUNE.

'Ye field-flowers, the gardens eclipse you, 'tis true ;
Yet, wildlings of nature, I dote upon you,
   For you waft me to summers of old,
When the earth teemed around me with fairy delight,
And when daisies and buttercups gladden'd my sight,
   Like treasures of silver and gold.'

UNE comes to us laden with such a bounteous, beauteous wealth of flowers, that it would require many a ramble in meadow, field, and wood to gather half of nature's treasures. We can only pluck those that first catch the eye, nor linger to discover the modest retreat of many a lovely blossom that hides amongst the grass or tangled hedge-way ; for

'The spirit of beauty is everywhere.'

Friends may boast their rich possessions in cultivated gardens and spacious conservatories ; but give me the

101

beautiful, fair woodlands wherein to wander, or the wild, open moor, where the fresh, health-laden breeze comes sweeping by. No gardener's leave is needed there, to gather the sweet flowers that are scattered in wild luxuriance all around, and which are

> 'A blessing given
> E'en to the poorest little one
> That wanders 'neath the vault of heaven.'

Spring has now quite given place to the warmer, richer days of summer; the hedgeways have assumed a fuller, the meadows a gayer, aspect than in the earlier months of the year.

> 'Retiring May to lovely June
>   Her latest garland now resigns;
> The banks with cuckoo-flowers are strewn,
>   The wood-walks blue with columbines;
> And with its reeds the wandering stream
> Reflects the flag-flower's golden beam.'

How fair and luxuriant are the woods, profuse in the flowering beauty of many a stately tree and more lowly shrub, the bright colouring of their various blossoms subdued and softened by the pure pale green or deeper tints of leafage. The beautiful white wax-like flowers of the glossy-leaved holly are now in bloom; the yellow-tinged blossoms of the snow-berry, with its pretty soft green leaf; the

> 'Laburnum, rich
> In streaming gold,'

droops its graceful pendent flowers, contrasting with the

dark green shrubs of rhododendrons, so profusely clad in their groups of delicately-tinged blossoms ; and

> ' The lilac, various in array, now white,
>   Now sanguine, and her beauteous head now set
>   With purple spikes pyramidal, as if
>   Studious of ornament, yet unresolved
>   Which hue she most approved, she chose them all.'

There is the beautiful guelder-rose, with its bunches of clustering white blossoms and fresh green leaves, which Cowper describes as

> ' Tall,
>   And throwing up into the darkest gloom
>   Of neighbouring cypress or more sable yew
>   Her silver globes, light as the foamy surf
>   That the wind severs from the broken wave !'

And oh, how delicious scents the

> ' Syringo, ivory pure,'

fair as the fairest lily, and yielding a perfume so powerful, that the very woods are fragrant !

> ' The sweet syringo, yielding but in scent
>   To the rich orange.'

It much resembles the orange blossom, and on that account is often called the mock orange. Yonder gleams the noble horse-chestnut, with its abundant fresh green leaves, and obtrusive spikes of showy white flowers. In less conspicuous beauty are the limes and sycamore, the bird-cherry, elder, etc. How beautiful are the hedges in these 'bonnie days of June,' with the pretty pink blos-

soms of sweet-brier, or poets' eglantine, entwined with
the fragrant honeysuckle !

> ' Yonder is a girl, who lingers
>   Where wild honeysuckle grows,
>   Mingled with the brier rose,
> And with eager, outstretched fingers,
>   Tiptoe standing, vainly tries
>   To reach the hedge-enveloped prize.'

The brier, or eglantine (*Rosa rubiginosa*), is crowded
with fragrant leaves and sharp prickles, its leaflets
covered beneath and at their margins with minute brown
glands, which give them such a sticky feeling.   But this

> ' Rain-scented eglantine,
> That gives such temperate sweets to that well-wooing sun,'

must be so common and familiar to you all, that description seems unnecessary.   Delicious is its perfume, borne
on the soft summer air after rain has fallen :

> ' O eglantine, sweet eglantine,
>   How rich a dower of scent is thine !
>   A perfume sweet unto the sense
>   As poetry's own eloquence ;
>   Of deepest, tenderest feeling born,
>   And bearing, e'en like thee, a thorn.'

Its berries are a beautiful scarlet, and much relished by
the birds.   Its blossoms are smaller, and of a deeper
pink than those of the common dog-rose (*Rosa canina*),
which is one of

> ' The sweetest flowers wild nature yields,'

BRIER OF EGLANTINE—*Rosa rubiginosa.*

COMMON DOG-ROSE—*Rosa canina.*

with its soft, faint odour, and delicately flushed petals, that so sweetly

'Doth peep forth with bashful modesty'

from amidst the strong, spiny stems, and luxuriant mass of rich green leaves. It is called dog-rose, from the belief that dogs relish the brilliant scarlet heps, that gleam so brightly in the autumn months, and which form so abundant a provision for birds. A conserve is also made from the fruit, and sold by druggists. There are no fewer than eighteen species of wild rose, some very common, amongst which is the Scotch or Burnet rose (*Rosa spinosissima*), which I dare say you have often observed blooming in heathy or chalky ground. Its leaves are much smaller, of a darker and less glossy green than those of the dog-rose; its flowers a soft, rich cream-colour, sometimes tinged with red, and its sharp, needle-like prickles crowd upon the stems.

There is a long, trailing, white dog-rose (*Rosa arvensis*), that has pretty white blossoms growing n clusters, and is especially common in Yorkshire. It is said to be the white rose chosen by the Yorkists as their badge in those terrible civil wars, of which you will all have read, commonly called the 'Wars of the Roses.'

The heps of all these roses are prized by birds, and yield them an abundant harvest. There are several pretty legends attached to the rose, which we are told was originally white, but received its beautiful blush tint from Eve, who, enraptured with its delicate beauty, pressed her lips on the snowy petals, and thus imparted a warmer tint of colour to the blossom. One of them further states that the stems were thornless until the expulsion from Paradise.

Mixed with this hedge of thorn and eglantine rise the beautiful white pyramidal clusters of the privet (*Ligustrum vulgare*), its abundant snowy blossoms rendering it a most elegant shrub.

> ' The privet, too,
> Whose white flowers rival the first drifts of snow
> On Grampia's piny hills.'

Its leaves are a rich, glossy green ; and its dark clusters of purple berries serve not only as winter food to birds, but yield a beautiful green dye.

Trailing through the hedgeway, we may find the pale pink blossoms of the blackberry (*Rubus fruticosus*), or the whiter flowers of the dewberry (*Rubus cæsius*), which has more pointed, longer leaves than the bramble. It is less common, too, and bears a bluish fruit.

The raspberry (*Rubus idæus*) is also common, especially so in Scotland, and forms a favourite covering for game. It has a woody, round, erect stem, and red fruit, which is considered superior in flavour to the cultivated plant, and which, like the blackberry, makes a delicious preserve.

> ' Ofttimes without man's aid we grow
> All independently, and throw

Scotch or Burnet Rose—*Rosa spinosissima.*

WHITE DOG-ROSE—*Rosa arvensis.*

> Our fruits upon the ground below,
> Where there is none to gather them.'

The mountain bramble or cloudberry (*Rubus chamce-morus*) blooms during this month on the summits of hills and mountains; and scarce has the snow dissolved from their cloudy tops ere it puts forth its pure white blossoms. It is an elegant plant. Its berries, which are a brownish-orange colour, ripen in August, and are preserved by the snow. The Highlanders and Laplanders highly esteem the fruit, which has an acid flavour. The scientific name of *rubus* is derived from the Celtic *rub*, signifying *red*.

Behold the pretty bright flowers of the soft, velvety silver-weed (*Potentilla anserina*), gleaming like stars from the hedge-side! The blossom is a pure yellow, the numerous leaves a pretty green, their under-surfaces wearing a grey, silvery hue:

> ' Silver-weed, with yellow flowers
> Half hidden by the leaf of grey.'

The roots of this plant are sometimes roasted like chestnuts, and eaten by village children.

And here beside it blooms a lovely cranesbill or geranium (*Geranium pratense*), which is not uncommon in the hedge during this month, but loves best to blossom in the moist copsewood. It is a tall plant, often as high as two feet; its leaves are deeply cut into segments; its delicate but handsome flowers a beautiful deep blue or purple colour, frequently as large as a halfpenny.

Here are some blossoms of the yellow goat's-beard (*Tragopogon pratensis*), which you all doubtless conclude is *only* a dandelion. But observe its long, tapering green leaves winding and twisting about its stems, and you will soon discover your mistake. It closes its flowers at mid-day, and, in consequence, is often called Jack-go-to-bed-at-noon.

And here also the

> ' Nightshade's purple flowers,
> Hanging so sleepily their turbaned heads,
> Rest upon the hedge.'

Let us gather a few sprays, and rest on this stile while we discuss them. First, we have the woody nightshade, or bitter-sweet (*Solanum dulcamara*), the garden nightshade (*Solanum nigrum*), and the common enchanter's nightshade (*Circea lutetiana*), which may all be abundantly found in the hedgeside during this month. The latter is a pretty little flower which we gathered in May, with spikes of delicate lilac, or pink-tinged blossoms, and elegantly-shaped soft green leaves. It grows about a foot high, loves best a shady place, and often grows wild in the neglected recesses of the garden. The garden nightshade is also often found in waste places, and flourishes on the sea-beach. Its flowers are white, its berries black, and, like all the nightshades, poisonous. It is a very common weed in our gardens, and to be found all over the world; and it is somewhat singular that in tropical countries its berries lose their deleterious qualities. The woody nightshade gains its name of bitter-sweet from the flavour of the root, which tastes

GROUP OF NIGHTSHADES.

bitter when first placed in the mouth, and afterwards becomes sweet.

> ' Shade-loving evergreen, say we not sooth
> When thee we liken unto fair truth?
> Bitter, full oft, is the draught from thy cup,
> But sweet is the taste which it leaves when drunk up.'

Its oval, brilliant scarlet berries, which during winter

hang in glistening clusters from the leafless branches, often tempt, by their beauty, young children to partake of their fatal, poisonous juices. Its flowers are composed of a yellow cone and purple petals, and, you will observe, greatly resemble the potato blossom, which in its botanical character is closely allied to the various species of poisonous nightshade ; but all narcotic and unwholesome effects are removed by the cooking the potato undergoes when prepared for food :

'The poor man's bread, the rich man's luxury.'

The potato was brought from America by Sir Walter Raleigh, who first planted it in his garden at Youghall, near Cork. The Spaniards called it *battata*, which we have corrupted into *potato*.

The deadly nightshade (*Atropa belladonna*) is a very different plant, much less common, but flowering also during this month. It may be found in the hedgeside ; but loves best an old ruin or quarry edge, where its branching stems will often grow four or five feet high. It bears handsome purple flowers, drooping and bell-shaped, which grow in pairs close to the large, dark green leaves. Its berries, which from green become a beautiful glossy black, and are nearly as large as a cherry, are highly poisonous, and it is said half of one is sufficient to prove fatal. History relates how the army of Sweno the Dane, when he invaded Scotland, fell victims to the faithless Scots, who, during a truce, mixed the berries of this nightshade in the drink of the Danes, and then murdered them when under the influence of the stupefy-ing sleep it produced. Its botanical name of *Atropa*

DEADLY NIGHTSHADE—*Atropa belladonna.*

belladonna is a somewhat incongruous and strange one. That of *atropa* has doubtless been bestowed on account of its poisonous properties, and is taken from Atropos, a powerful goddess, one of the Parcæ or Fates, three sisters, whom the ancients supposed to preside over the birth and life of mankind. Atropos was the eldest, and believed to cut the thread of human life with a pair of scissors. The other name of *belladonna*, or *fine lady*, is given in consequence of the use made of the plant by Italian ladies to remove pimples from the skin; and many in our own land apply a wash made from it for beautifying their complexions. As a remedy in some cases of disease in the eye, it is, I believe, most valuable.

But now we have rested, let us mount this stile leading into the corn-fields, and look down on their many revealed beauties.

Waving so gracefully to and fro 'neath the summer breeze is the common scarlet poppy (*Papaver rhœas*), so gay in its brilliant livery, and rich, deep purple eye:

> ' The poppies red,
> On their wistful bed,
> Turned up their dark blue eyes for thee.'

This fragile, ill-smelling, finger-staining flower is often called headache; and by this name Clare speaks of it:

> ' Corn-poppies, that in crimson dwell,
> Called headaches, from their sickly smell.'

We have six species of wild poppy, amongst which is the graceful white poppy (*Papaver somniferum*), which may also be found in our fields. It is very extensively cultivated, both in European countries and India, for the excellent oil that is extracted from its seeds, and the opium, so valuable as a medicine and narcotic.

'O gentle sleep,
Scatter thy drowsiest poppies from
above.'

I have read that, in the East, poppy - seeds are sprinkled on the tops of cakes; and that the cracknels which Jeroboam sent to the prophet Ahijah, when he inquired the fate of his sick child, are supposed by Dr. Kitto to have been cakes sprinkled in this way. From this may, perhaps, have originated our own custom of adorning shortbread and other cakes with caraway seeds and comfits.

Showy and beautiful as we find 'the flaunting poppy,' the farmer views it with no such admiring eye; for he looks on his fields, and sees

'There poppies nodding, mock the
hope of toil.'

And the same may be said

Corn-Cockle—*Agrostemma githago.*

of the corn-cockle (*Agrostemma githago*), sometimes called 'crown of the field,' of which we have but one British species. It is a tall, erect plant, the flower finely formed, and having the fine long green leaves of the calyx standing boldly out round the petals. Its stems and leaves are somewhat hairy; but its rich purple blossoms render it highly ornamental.

The scarlet pimpernel (*Anagallis arvensis*) grows in the borders of our corn-fields, and, with the exception of the poppy, is our only scarlet wild-flower. It is called 'the poor man's weatherglass,' on account of its closing its brilliant blossoms in timely notice of rain.

> 'When with clouds the heavens frown,
>   Then thy little head bends down.
>   Little weather-prophet, say,
>   Fair or foul the coming day?
>   Little scarlet pimpernel,
>   None can tell us half so well
>   What the coming change shall be,—
>   None but such an one as thee.'

About noon, however, it closes its petals for the day, regardless of the brightest sunshine. There is not a prettier flower than this 'little scarlet pimpernel,' with its bright, merry face, and blue or violet coloured mouth, with tufts of gold-tipped stamens, its delicate, trailing stems, and dotted leaves. Take heed how you gather it, for nettles abound,—nettles in which, I am sure, my young friends see no beauty, and are afraid to touch; but *grip* them *firmly*, and they will harm you not:

' If you gently stroke a nettle,
　　Mark, it stings you for your pains ;
　　But seize it like a man of mettle,
　　And it soft as silk remains.'

All do not sting like the small nettle (*Urtica urens*), with green blossoms, or the great nettle (*Urtica dioica*), with still brighter green flowers, sometimes tinged with purple or red. Mary Howitt says :

' The nettle he throve, and the nettle he grew,
　　And the strength of the earth around him he drew.'

And true indeed is this, applied to our great nettle, which grows by every wayside, often as high as three feet :

' The stuff of his leaves is strong and stout,
　　And the points of his stinging flowers stand out,
　　While the child that runs 'mid the grass to play,
　　From the great king-nettle will shrink away.'

We have, however, a third species, even more virulent, but happily less common. This is the Roman nettle (*Urtica pilulifera*), which is usually seen by the seaside. The botanical term *urtica* is derived from *uro*, to burn.

The white dead nettle (*Lamium album*) is rather a handsome plant, with its large leaf, and white flowers slightly touched with pink, growing in whorls around the stem. Bees and butterflies love it well. It is called *dead*, because possessing no stinging powers. Our common name of *nettle* is supposed to be derived from the Anglo-Saxon word *nædl* or *netel*, a needle.

Despised as the nettle generally is, only to be found

in neglected spots, boasting neither beauty nor sweet fragrance, it is nevertheless a useful plant, possessed of many valuable properties. The peasant esteems it as a safe medicine, and often brews his 'nettle tea.' The young shoots, when boiled, are said to resemble asparagus. In times of famine it has been known to sustain the poor for days. The stalks of the old plant may be dressed after the manner of flax, and woven into cloth. Green and yellow dyes may be extracted from its juices. Few beasts will eat the *growing* nettle, though they seem to relish it when dried, and mixed with their fodder; and nettles are often boiled for pigs' meat. Hosts of our beautiful winged insects find shelter and food in the nettle; and as surely as we find it growing by every roadside, so shall we see it adorned by a bevy of fluttering, gorgeous butterflies.

And now, leaving the corn-fields, we wander down to the streamlet's side to gather a bunch of the true forget-me-not, or water-scorpion grass (*Myosotis palustris*), with its beautiful blue petals, and little yellow eye, gleaming so brightly. There is an old legend, which relates how this flower was named by a drowning knight, who, flinging it from the stream on to the bank at the feet of his lady-love, exclaimed, 'Forget me not!' and sank to rise no more. But Agnes Strickland tells us that it was Henry of Lancaster who made it the symbol of remembrance. And thus, like the broom of the Plantagenets, the fleur-de-lis of Henry VII., and the hawthorn of the

Tudors, it became an historical flower. There are eight species of this plant. In the wood we may gather the wood-scorpion grass (*Myosotis sylvatica*); and in the hedge-bank the pretty turquoise blossoms of the field-scorpion grass (*Myosotis arvensis*), so often mistaken for the *true* forget-me-not. All are blue, and all lovely.

But a more beautiful ornament to the still stream than even the forget-me-not, is the water-violet or feather foil (*Hottonia palustris*). Its pale lilac flowers, growing in whorls around the stem, rise above the stream, but its leaves are all submersed. It resembles the cuckoo-flower much more than the violet. Its scientific name was bestowed in compliment to Hotton, an eminent botanist.

By the water's edge blooms the water figwort or water betony (*Scrophularia aquatica*),—a very troublesome plant, I believe, to anglers. It has small, dingy, greenish-purple blossoms, with notched, pointed leaves, heart-shaped at the base, and square stem. The knotted figwort (*Scrophularia nodosa*), which we shall now find in the wood, or shady place, greatly resembles it, though a taller plant, its small flowers growing in looser clusters, and its stem wanting the green expansion on the square edges, so peculiar in the water figwort.

Beautiful and fragrant is the meadow-sweet (*Spiræa ulmaria*), or 'queen of the meadow,' as it has been called, growing some three feet high, and adding grace and beauty to the landscape, with its delicate waving clusters of many small blossoms, as pure and white as though formed of newly fallen snow. Its leaves are handsome and jagged, its stems tinted with red, and its delicious odour resembles hawthorn and freshly mown hay. Its 'foam - like flowers' often deck the stream side. It loves best a damp soil :

> 'In the moistened plain
> The meadow-sweet its lusci-
> ous fragrance yields.'

WATER-AVENS—*Geum rivale.*

Another friend we may find here, the marsh cinquefoil (*Comarum palustre*), which has a dingy purple blossom, growing on an upright, branching stem ; and its frequent companion, the water-avens (*Geum rivale*), with its graceful drooping flowers, a dull, purplish orange, marked with dark veins.

The valerian (*Valeriana vernus*) also loves the moist

meadow, and is very abundant during this month. It is a pretty pale blossom, with clustering, pinky-white flowers, and tall, beautiful leaves. It is said of it:

'Accommodating plant, that grow'st in moist or arid ground,
  Thou flourishing, alike in bleak and sheltered spots, art found.'

But yonder rise the handsome blossoms and large glossy green leaves of the yellow water-lily (*Nuphar luteum*), with its thick stems and beautiful leaves, on which rests its rich yellow cup, with all the dignity of a queen among her maidens. Exquisite as it is, however, it is not so beautiful as the white water-lily (*Nymphæa alba*), which blossoms in August:

'The white water-lily, so wondrous fair,
  With her large broad leaves on the stream afloat,
  Each one a capacious fairy boat.'

These elegant flowers lift their heads in early morning, and gradually expand their beauties to the rising sun. All day they proudly bask in his genial warmth, but as the evening shadows lengthen, they softly sink to rest, sometimes reposing on the bosom of the water, at others stealing beneath its surface, until the returning dawn bids them once more unfold their silver and golden blossoms. Our yellow lily is sometimes called *brandy bottle*, from its peculiar scent, a somewhat unromantic name for

'Those lovely lilies, that all the night
  Are bathing their beauties in the lake,
  That they may rise more fresh and bright,
  When their beloved sun's awake.'

But these beauteous treasures are beyond our reach to gather for our bouquet; therefore let us away to the green fields and banks, where we shall find nature equally profuse.,

How prettily is the wayside decked with the soft green leaves and velvety yellow blossoms of the creeping cinquefoil (*Potentilla reptans*), that creeps along the ground, its glowing, starry face upturned to the summer sun! And brightly shining amongst the grass is the little butterfly-shaped bird's-foot trefoil (*Lotus corniculatus*), familiarly called pattens-and-clogs, which is often sown in the field to form pasture for cattle.

Here, too, the sweet-scented common agrimony (*Agrimonia eupatoria*), so prized by the village herbalist, raises its long spike of yellow flowers

BIRD'S-FOOT TREFOIL—*Lotus corniculatus.*

and deeply notched leaflets, so varied in size; as well as the tall, very pretty bladder campion or catchfly (*Silene inflata*). Its blossoms are white, its leaves sea-green; and examine how beautifully its inflated flower-cup is veined as with network! Ancient writers inform us that this blossom, named after the god Silenus, was originally a youth named Campion, who was employed to catch flies for Minerva's owls to feed upon during the day, when they lacked the sight to hunt prey for themselves. The unfortunate youth happening to fall asleep in the midst of his occupation, the infuriated goddess changed him into this flower, which still bears the form of the bladder or bag in which Campion secured the flies, and droops its timid head as night approaches, to escape the sight of the owls that then wing their flight in the air.

The short-lived, sweet-scented field convolvulus, or bindweed (*Convolvulus arvensis*), is wreathing its delicate pink blossoms on hedge-bank and field,—a troublesome plant to the farmer, but one well loved by the bee,

> ' Flying solicitous from flower to flower,
>   Tasting each sweet that dwells
>   Within its scented bells.'

It is a pretty, fragile flower, that closes at the approach of rain or evening:

> ' The bindweed frail, that, when the light
>   Departs, aye bids the world good-night;
>   And foldeth up its silken vest,
>   As though intent on seeking rest.'

Lovely as this blossom is, it is exceeded in beauty by the great bindweed (*Calystegia sepium*), which in autumn adorns our fading hedges with its large snow-white bells.

BINDWEEDS—*Convolvulus arvensis; Calystegia sepium.*

We have scarcely any wild-flower more elegant in form, or purer in tint. Unlike the smaller convolvulus, however, it possesses no scent, but its blossom is equally fragile. It loves a somewhat moist situation, and is a great pest to the agriculturist.

We have also a third convolvulus, called the sea bind-weed (*Calystegia soldanella*), which decks our coasts with its trailing clusters of large pink yellow-rayed flowers, and dark green succulent leaves. This and the great bind-weed formerly belonged to the genus convolvulus, de-rived from the Latin *convolvo*, on account of its twining character. *Calystegia* is taken from two Greek words signifying *beautiful* and *a covering*, in allusion to the bracteas which distinguish these bindweeds from the true convolvulus. Perhaps I ought here to explain that the bracteas are those floral leaves which expand them-selves at the same time as the flower, and differ from the leaves in general, as well in colour as form.

At the seaside we might now also gather the pale yellow blossoms of the sea-cabbage (*Brassica oleracea*), which ornaments the chalky heights with its abundant clusters of handsome flowers. It is the origin of all our cultivated cabbages, which are scarcely recognisable in this wild, straggling, heartless plant.

And almost as common on our English coast is the sea-holly (*Eryngium maritimum*), with its sea-green, spiny leaves, delicately veined with white, its large, tough stem, and blue, thistle-like flowers, which burst in masses from their scaly receptacle. It is a sturdy plant, blooming freely during this and the two following months, in defiance of weather or exposed situation.

' Eryngo, to the threat'ning storm
    With dauntless pride uprears
His azure crest and warrior form,
    And points his spears.'

The sea-pea (*Lathyrus pisiformis*) flowers now,
spreading its clusters of purple blossoms over the sandy
grass ; as well as the sea-rocket (*Cakile maritima*), with
its pretty pink-white flowers, and soft bluish-green leaves ;
the scurvy grass (*Cochlearia maritima*), with white flowers,
and tufts of kidney-shaped leaves; and the rock sam-
phire (*Crithmum maritimum*), with its pale green, fleshy
leaves, which make so delicate a pickle for our table.

But no blossom on our coasts is so gay as the beautiful
sea-celandine, or yellow-horned poppy (*Glaucium luteum*),
which grows close down by the sea, drooping its fragile
petals with every whisper of the breeze. Its flowers are
a brilliant amber colour, glowing in low bushy groups of
delicate sea-green leaves, which, with the stem, are covered
with that peculiar bloom called by botanists glaucous.
Its peculiar feature is the curved seed-pod, that grows
boldly out, often half a foot long.

But this is a digression from the blossoms of our country
ramble, where in the pasture or hedge-side we shall now
find blooming the common yellow rattle (*Rhinanthus
Crista Galli*), so tall and erect, with its small yellow blossoms and swollen flower-cup. It belongs to the figwort
tribe, grows about a foot high, and is called rattle, from
the noise made by every passing breeze shaking the ripe

YELLOW-HORNED POPPY—*Glaucium luteum.*

seeds in their husks. This plant is a great annoyance to the husbandman; and so is that pretty pea-shaped blossom, the rest-harrow (*Ononis arvensis*), which has earned its name from the way its tough roots and branches often check the action of the harrow. It is often very plentiful on the links and chalky cliffs of the seashore. Its flowers are a purplish pink, sometimes white, and it is armed with many strong spines.

Near the rest-harrow, the common juniper (*Juniperus communis*), so extensively used in the flavour of gin, is now in bloom. In England it never attains any size; but in other countries it rises to the height of trees, and often affords a green and fragrant shade. And from this you can understand how Elijah, when threatened by Jezebel, and fleeing to the wilderness, 'lay and slept under a juniper-tree :'

> ' When from King Ahab's wrath Elijah fled,
>   This mountain shrub gave shelter and a bed.'

The Swedes convert its berries into a preserve and beer. In Norway its branches are strewn on graves.

Many of the orchis tribe are now in bloom:

> ' In the greenwood, on the hill,
>   ، Fluttering to the breeze,
>   Gay as the fly that's never still,
>   Grow the orchises.'

In almost any hedgeway or wood we shall find the ، beautiful purple orchis (*Orchis mascula*) of early summer, which has rich purple blossoms marked with dark spots, and grows on a tall, succulent stem, around which the leaves gather. It is strongly scented with an unpleasant odour. The green-winged meadow orchis (*Orchis morio*), on the contrary, is often sweetly fragrant. Its pinkish-purple blossoms grow thickly amongst the grass. The largest orchis of all is the lady orchis (*Orchis fusca*), which often grows as high as two feet. Its flowers are of a brown purple, and grow in large clusters. The birds'-nest orchis (*Ophrys nidus-avis*) looks most like a withered leaf fallen to the ground. The bee orchis (*Ophrys apifera*) and fly orchis (*Ophrys muscifera*) bear a strong resemblance to those insects; and one is apt at first to suppose they have actually alighted on the blossoms:

> ' I sought the living bee to find,
>   And found the picture of a bee.'

The flowers of the bee orchis are brown, variegated with yellow, like the body of the bee. The fly orchis is smaller. There are many other varieties, blooming throughout the summer months, when

> ' Each dry entangled copse empurpled glows
>   With orchis blooms.'

But I fear we cannot consider them to-day. The root of the orchis, after flowering, produces a bulb from its side, and then dies off, the new root strengthening and becoming the flowering root of the succeeding year.

Decking the waste places of our land are the brilliant blue blossoms of the common borage (*Borago officinalis*). Its flowers are large, with prominent stamens, its stems and leaves covered with prickly hairs. The bees hover incessantly over it, with busy wing and melodious hum.

COMMON CISTUS or ROCK ROSE—*Helianthemum vulgare.*

It was said the flowers steeped in wine proved an invigorating cordial:

> ' Famed for driving care away,
> Rough borage.'

Also here we have the common cistus, or rock rose (*Helianthemum vulgare*), the pink variety of which is supposed by some to be the rose of Sharon mentioned in Scripture. The crumpled petals are very delicate, yet bright in their golden beauty, adorning many a barren spot with clumps of golden flowers. They fade rapidly, but are so abundant, that they follow each other in quick succession :

> ' The little rock rose, oh ! it fades in a day,
> As popular favour passes away.'

How beautifully is the brilliant red saintfoin (*Hedysarum onobrychis*) gleaming under the sunlight rays! It loves a chalky soil, and decks many a hedge-bank ; it is also cultivated in the field for food for cattle.

But we are neglecting to gather one of the handsomest flowers of our hedgeside, the tall, showy foxglove (*Digitalis purpurea*), which

> ' Rears its pyramid of bells,
> Gloriously freckled, purpled, and white.'

It is indeed one of the most splendid blossoms we have, raising its commanding spike of flowers with such stately grace by the borders of our woods and forests.

> ' The foxglove tall
> Sheds its loose purple bells, or in the gust,
> Or when it bends beneath the upspringing lark,
> Or mountain finch alighting.'

It often grows to the height of four or five feet, the numerous blossoms occupying about half the length of the stem. Its leaves are a dull green, large, and veiny.

Let us
> ' Explore the foxglove's freckled bell,'

and we shall find that its tubular flowers,

> ' In whose drooping cups the bee
> Makes her sweet music,'

sometimes purple, sometimes cream-coloured or white, are beautifully speckled within.

Its name is said to be a corruption of fairies' or folk's love, because it was supposed that fairies loved to lurk in its bells ; and many superstitions were connected with it, in consequence of its being held a sacred plant by the Druids, who used it in their midsummer sacrifices. It possesses poisonous properties ; but is highly serviceable as a medicine, and seems to be dearly loved by the

> ' Bees that soar for bloom
> High as the highest peak of Furness Fells,
> And murmur by the hour in foxglove bells.'

As we pass the bare wall, we may pluck a patch of the pretty yellow blossoms of the biting stonecrop (*Sedum acre*), or wall-pepper, as it is sometimes called from its acrid juiciness ; and, before concluding our ramble, gather a few flowers from the various useful flax plants :

> ' Oh, the goodly flax-flower !
> It groweth on the hill ;
> And to the breeze, awake or sleep,
> It never standeth still.

It seemeth all astir with life,
  As if it loved to thrive ;
As if it had a merry heart
  Within its stem alive !

Then fair befall the flax-field ;
  And may the kindly showers
Give strength unto its shining stems,
  Give seed unto its flowers.'

In this month the pretty little cathartic flax (*Linum*

catharticum) is in bloom, and abundant by the roadside, though not conspicuous, for it grows only a few inches high, and scarcely raises its slender stems, laden with drooping buds and small erect white blossoms, above the summer grass. So delicate and beautiful, however, are its small silvery flowers, that it has been called the 'fairies' flax.'

We may also find the beautiful blue blossoms of the perennial flax (*Linum perenne*), and the paler blue kind, called, from its tapering leaf, the narrow-leaved pale flax (*Linum angustifolium*), which has lighter but perhaps purer blue flowers. Both may be found adorning the chalky hillside, or sandy pasture of the sea-coast, and are alike provokingly fragile, scattering their petals at

CATHARTIC FLAX
—*Linum catharticum.* every attempt to gather them. But the

species most cultivated for commerce on account of its fibrous stems, from which our valuable linen material is made, and which Mrs. Howitt so truthfully describes in our quoted verses, is the common flax (*Linum usitatissimum*). Often will you find it springing up in the corn-fields; and truly beautiful are its silky blossoms, with their slender stems and tapering green leaves, that bend and wave with such delicate grace 'neath every passing breeze. It grows about two feet high, and its large erect blue flowers are richly marked with purple veins. There is a peculiar exquisite delicacy in this azure-blossomed flax; and very prettily has Coleridge spoken of

NARROW-LEAVED PALE FLAX—*Linum angustifolium.*

'The unripe flax;
When, through its half-transparent stalk at eve,
The level sunshine glimmers with green light.'

We read of the cultivation of the flax from the earliest ages; and in several parts of the Bible it is mentioned. In former times our forefathers reared it in sufficient quantity to supply their families with linen, the ladies of the household spinning and weaving the yarn. The flax is one of our most valuable and in-teresting plants, not only on account of its fibrous stem,

COMMON FLAX
—*Linum usitatissimum.*

but for the many uses of the seed, which succeeds the flowers, each little round pod being filled with about ten of those small flat brown seeds called linseed, which I think must be so familiar to you all, so much used for poultices, and from which a valuable oil is expressed, the refuse being given to feed cattle or manure the land.

It is in the stems of the flax that the fibres are found which compose our linen ; and when these are required, the plant is gathered immediately after flowering, being carefully pulled up by the root and dried in the sun. It is then steeped in water, or left to lie in a tank or shallow river until the outer covering of the stalk decays, when it is spread out to dry and bleach, and the fibres are afterwards easily separated from the pith and skin. They are next combed with iron combs, and formed into the tow, which is first spun into yarn, and afterwards woven into cloth. For the manufacture of fine laces, lawns, etc., much combing is required, and only the finer threads taken. The scientific name of *Linum* signifies *thread,* and is taken from the Celtic word *lin.* The cultivation of flax being very exhaustive to the soil, is not much favoured in England; but it is largely grown in France, Holland, Russia, Egypt, India, etc. Ireland produces nearly all the flax used in its extensive manufacture of linen; and in Scotland, as well as a few of the midland counties of England, many a field of flax waves in the summer beauty of its sweetly blue blossoms.

'Ah, 'tis a goodly little thing,
    It groweth for the poor,
And many a peasant blesses it,
    Beside his cottage door.

He thinketh how those slender stems,
    That shimmer in the sun,
Are rich for him in web and woof,
    And shortly shall be spun.

He thinketh how those tender flowers,
    Of seed will yield him store;
And sees in thought his next year's crop,
    Blue, shining round his door.

Then fair befall the flax-field;
    And may the kindly showers
Give strength unto its shining stems,
    Give seed unto its flowers!'

We can never sufficiently appreciate the value of this

plant, nor the loving-kindness of the beneficent Creator who has so richly provided for our every want. 'He causeth the grass to grow for the cattle, and herb for the service of man : that he may bring forth food out of the earth.' 'Bread which strengtheneth man's heart.' He hath said, 'While the earth remaineth, seed-time and harvest shall not cease ;' and every season bears witness to the fulfilment of his promise.

> ' He giveth to the beast his food,
> And to the young ravens which cry.'

He drops sweets into the beauteous flowers to sustain the bee and insect tribes ; with the silky down of the thistle, and wild berries that glimmer and grace the hedgeway, He feedeth the happy birds ; and by the grass of the field He nourisheth 'the cattle upon a thousand hills.'

VII.

JULY.

' Everywhere about us are they glowing ;
    Some, like stars, to tell us spring is born ;
Others, their blue eyes with tears o'erflowing,
    Stand, like Ruth, amid the golden corn.'

RIEF though the time has been since we enjoyed our last sunny ramble together, a great change has come over the spirit of earth's dream. The 'merry spring' blossoms have all faded and gone ; the fresh beauty of the year has passed away ; the fair green of the forest leaves has given place to deeper tints and denser foliage ; the light-hearted birds at noon-day are silent in the woods ; and the hot July sun of full summer shines gloriously upon us.

' The primrose to the grave is gone,
    The hawthorn flower is dead ;
The violet, by the mossed green stone,
    Hath laid her weary head.'

141

Yet still the earth teems with profusion fresh ; new beauties are springing all around. We will not grieve that spring has departed, delicate and lovely as she is, when the richer, more luxuriant beauty of summer fills her place.

> ' They may boast of the spring-time when flowers are the fairest,
>   And birds sing by thousands on every green tree ;
> They may call it the loveliest, the greenest, the rarest,
>   But the summer's the season that's dearest to me.
>
> Yes, the summer, the radiant summer is fairest,
>   For green-woods, and mountains, for meadows and bowers,
> For waters, and fruits, and for flowers the rarest,
>   And for bright shining butterflies, lovely as flowers.'

Beautiful are now the exquisite drooping bunches of fair acacia, so graceful in their waving, snowy purity, or

> ' White with faintest crimson flush,'

and bunches of elegant green leaves. How sweet the faint odour of the lime-trees blossom, that hangs its pretty sprays of yellowish-white bells, beneath their sheath of softest green ! Flowers and leaves are spread down to the very margin of the rivulet ; and even on its limpid surface are

> ' Green tufted islands, casting their soft shades
>   Across the burn-sequestered leafy glades.'

What can be more beautiful in nature than our handsome water-lilies, with their rose-like blossoms, and large, smooth, oval, green leaves, resting on the top of the water,

> ' Making the current, forced awhile to stay,
>   Murmur, and bubble as it shoots away?'

or the pretty little flower of the water arrow-head
(*Sagittaria sagittifolia*), which is now in bloom, and
easily recognisable by its white blossoms and arrow-
shaped leaves, which lie in masses on the surface of the
stream ?   Whilst by the water's edge,

> ' Rich cymes of fragrant meadow-sweet '

wave their creamy clusters, 'neath every passing breeze,
and

> ' Where the still streamlet wanders o'er the glade,
>   The pungent cresses grow.'

There is not a more refreshing or wholesome salad than
the common water-cress (*Nasturtium officinale*); only we
must gather it cautiously, for it somewhat resembles the
water parsnip (*Sium nodiflorum*), a highly poisonous
plant near which it often grows. The blossoms of the
water-cress are small, white, and cross-shaped. It is
much cultivated in streams near large towns ; and many
poor persons make their living by gathering and selling
it. In France it is dressed as spinach, and forms an
excellent dish.

I have read somewhere,

> ' It was an ancient saying and belief,
>   That those who oft partook of cresses green,
>   Straightway became of public men the chief,
>   Of purpose firm, and resolute of mien.'

The great water plantain (*Alisma plantago*) is also a great ornament of the pond, with its large, handsome leaves rising on long stalks from the root, and small, delicate, lilac flowers composed of three petals, blossoming on its many-branched stems; as is also the common frogbit (*Hydrocharis morsus-ranæ*), its fragile, and also

COMMON FROGBIT—*Hydrocharis morsus-ranæ.*

three-petalled white blossoms, and thick glossy leaves floating in large patches on the water. Its stems lie horizontally on the surface; but I have slightly raised them in my illustration, that you might the better observe the sheaths from which the flowers spring, and the curious way in which the young buds are enfolded in their casing, through which they may be distinctly discerned.

On the moist bog we may find the bog-pimpernel (*Anagallis tenella*), with its tiny oval leaves, its delicate, slender stalks, and sweet little pink blossoms; and

> ' On the pool's half dry banks, there the red and green hue
> Of that small moorland darling, the little sundew ;
> Each plant lying close, like a 'broidered rosette,
> Shining redly with ruby-gems thick o'er it set.'

This small but singular plant, the funny little sundew (*Drosera rotundifolia*), has white blossoms and crimson-tinged leaves, thickly covered with hairs, tipped with a clammy liquid, which give it the appearance of being covered with the evening dew. Insects, attracted to the sweet fluid, become entangled in these clammy hairs, are thus captured, and die.

But this is the month of beauty to the wild moorlands, that in winter seem so drear and barren :

> ' The tiny heath-flowers now begin to blow ;
> The russet moor assumes a richer glow ;
> The powdery bells, that glance in purple bloom,
> Fling from their scented cups a sweet perfume.'

' E'en the slight harebell,' on its slender stem, bends its azure head in grace and beauty, stirring to every faint whisper of the passing breeze. Come then, hasten, my young friends, and let us away to the beautiful moor, so

K

rich in its array of flowers, where the butterfly and the bee are

> ' Feeding upon their pleasures bounteously.'

The commonest beauty of the heath is the common ling, or heath (*Calluna vulgaris*), a low-tufted shrub, with small leaves, and tiny lilac blossoms, which is included amongst the five species of heath, or the more general name of heather. They all have lovely flowers, varying from a dark purplish-red to a delicate rose-colour, and even white. The commonest of all, with the exception of the ling, is the fine-leaved heath (*Erica cinerea*). Its flowers are larger, and hang like delicate pink bells from the small, fine-leaved stems. The common ling is used for a variety of purposes in the northern counties. The Highlanders make their beds of its sprays, and use the coarser branches to make brooms, thatch their huts, and assist, with a kind of cement, in the construction of walls. It is burned for winter fuel; the fibres of its stems are twisted into ropes; and in the Western Isles it affords a yellow dye. It shelters and supplies with food many of our feathered tribe, especially grouse. It nourishes many a bright-winged insect, and

> ' Here their delicious task the fervent bees,
>   In swarming millions tend.'

In Scotland many of the heaths have been adopted as the badges of the different Highland clans.

Mixed with these shrubs of ' blooming heather,' that in their flush of beauty give so rich a purple colouring

to the hills, is a tiny blossom likely to be overlooked by
a careless observer. It is the pretty little common eye-
bright (*Euphrasia officinalis*), that may be found studding
the sides of the chalky hills, or half lost in the hedge-side
or tall grass of the fields, and which has received its
botanical name from a Greek word signifying joy or
pleasure. It is a cheerful, lowly thing ; its white blossoms sometimes varying to a delicate lilac, streaked
with darker lines ; its leaves a bright green and deeply
notched. It is usually only about two inches high ; but
varies according to the situation of its growth. It has
received its English name from its supposed efficacy in
curing diseases of the eye, and Milton speaks of

> ' Euphrasy, to cleanse the visual ray.'

But modern oculists consider it rather injurious than
beneficial. It is a sweet little blossom, however, the
sight of which must ever gladden the eye, as poet
Elliott must also have thought when he wrote the
verse :

> ' Sweet eyebright ! loveliest flower of all that grow
> In flower-loved England ! Flower whose hedgeside gaze
> Is like an infant's ! What heart doth not know
> Thee, clustered smiler of the bank ?'

And now we come upon

> ' A bank,
> Whereon the wild thyme blows.'

A lowly, spreading plant, of rapid growth :

> ' The humble, creeping thyme,
> Which, with a thousand roots, curling and crisp,
> Goes decking the green earth with drapery.'

Delicious, as we tread upon it, is the scent that rises from its small purple flowers and fragrant leaves.

> ' Thyme, the love of bees, perfumes the air.'

The wild thyme (*Thymus serpyllum*) is often infused as a cure for headache amongst country people ; and a strong oil is extracted from it. And here,

> ' With thyme, strong-scented, 'neath one's feet,
> Are marjoram beds, so doubly sweet.'

The sweet marjoram (*Origanum vulgare*) loves a dry, hilly bank, and will often grow a foot high :

> ' Sweet marjoram, with red-tinged leaves like blushes,
> Scattered amid the hillside bushes.'

Like the thyme, it yields an essential oil, which is often applied for the cure of toothache and rheumatism. Its blossoms, growing in clusters, of a deep purple hue, are aromatic and sweet. I know of only one British species. Both the marjoram and thyme are cultivated in kitchen gardens.

But there is a flower we failed to gather in our last ramble, which still covers the dry hillside,

> ' Where sweet air stirs
> Blue harebells lightly, and where prickly furze
> Buds lavish gold.'

It is the broom (*Cytisus scoparius*), 'the bonnie, bonnie broom,'—an evergreen now covered with myriads of golden blossoms, of which Wordsworth writes :

> ' On me such bounty summer pours,
>   That I am covered o'er with flowers ;
>   And when the frost is in the sky,
>   My branches are so fresh and gay,
>   That you might look at me and say,
>   That plant can never die.
>   The butterfly, all green and gold,
>   To me hath often flown,
>   Here in my blossoms to behold
>   Wings lovely as his own.'

It is not only the delight of butterflies and bees, which hover incessantly about it, but is made serviceable to man. Its branches are used in tanning leather ; its fibres may be manufactured into a coarse cloth ; the young buds are pickled in vinegar, and eaten as capers ; the wood, when old, is valuable to the cabinetmaker ; and the early shoots are often mixed with hops in brewing. It is used for thatching cottages, for making fences and brooms ; and, in Scotland, converted into fuel for winter firing. The flower resembles the furze, both in shape and colour ; its leaves are small and few, and its branches spring up erect and stiff. The French call it 'le genêt ;' and it is said the name of Plantagenet was derived from this flower. The first Earl, as a penance for some wicked action, allowed himself to be scourged with broom-twigs ; and ever after took the

name of Plantagenet, or *broom plant.* It is considered
the emblem of humility :

> ' A pilgrim to the Holy Land, Fulke, Earl of Anjou, went,
> Enjoined to expiate some crime in toilsome banishment ;
> Placed in the scalloped hat, he wore a sprig of lowly broom,
> And hence we name humility, the plant of golden bloom.'

We must not, however, linger on the moors, unless
it be to gather a handful of the beautiful blue harebell
(*Campanula rotundifolia*), that dances amongst the
prickly furze, nodding and bowing with every passing
breeze that plays upon the hill-side. Few of you can
fail to recognise the beautiful harebell, the blue-bell of
Scotland, which, I think, must be as familiar to you all
as is the equally beautiful hyacinth, or English blue-
bell, which we gathered in April. It blooms on our
commonest roadside, and peeps from out every dry
pasture or hedgeway ; but it is on hills and moors it
loves best to droop its pendent blossoms in greatest
and wildest luxuriance. Scott speaks of it as

> ' The little flower that loves the lea ;'

but Miss Twamley tells us,

> ' No rock is too high, no vale too low,
> · For its fragile and tremulous form to grow ;
> It crowns the mountain
> With azure bells,
> And decks the fountain
> In forest dells.
> It wreaths the ruin with clusters grey,
> Nodding and laughing the live-long day.'

HAREBELL—*Campanula rotundifolia.*

151

Ah! it is a lovely little blossom,

> ' That daintily bends its honeyed bells,
>   While the gossiping bee her story tells.'

As lovely in purity of colour as it is graceful in form, its delicate head ofttimes bowed to the very ground, *bent* by the sweeping blast, but not *broken*, owing to the little resistance offered by its narrow leaves, and the wiry elasticity of its stems. Its scientific name —*Campanula rotundifolia*—signifies *round*-leaved *bell*-flower ; and you will wonder to hear it so named, when its leaves, shooting out from the hair-like stems, are all so long and narrow ; but examine those growing near to the root, and you will find them round, or heart-shaped. These soon, however, wither and die off, as the plant increases in size and age. Its stem is some-times divided into many branches, but oftener simple, each bearing its own little blossom at the summit. It well merits its name of *blue-bell*, for few flowers can boast so pure an azure tint, or are more bell-like in shape. I have also heard it called heath-bell and fairy-bell ; the latter in reference, doubtless, to the pretty legend that is attached to it, of how, when the fairies . held their midnight revels,

> ' And glow-worm lamps illum'd the scene,'

it rang its chimes of fairy music,

> ' Perchance to soothe the fairy-queen,
>   With faint sweet tones, on night serene,
>   Its soft bells pealing.'

We are told that, 'once upon a time,' a shepherd boy, tending his flock on the hill-side,

> 'Where harebells blossomed wild and free,'

happened to lead them into one of those magic circles where the fairies had been revelling the eve before ; and as the sheep, wandering amongst the blossoms, chanced to touch them with their feet, each tiny bell commenced to ring its little chime of soft melodious sound, which so charmed both boy and flock that, entranced, they stayed to listen, and lingered on, forgetful of the passing hours. Shades of evening at length beginning to fall, the shepherd was reminded it was time to wend his steps homeward ; but in vain he strove to lead his charge away from that harmonious music. They would not stir beyond the limits of the magic circle ; for hark !

> 'Whene'er the little airs arise,
> How the merry blue-bell rings
> To the mosses underneath !'

He tried with gentle love to coax them forward ; but every persuasive art failed. Impatient at their obduracy, he attempted to *drive* them on ; but all to no purpose. At length, in the midst of his despair, when evening was yielding the world to deepening night, a happy thought occurred to his mind ; and stooping, he gathered a bunch of the delicate blossoms, which, in consequence of his rough and hasty handling, rang

out a louder, more melodious peal than before; when immediately the flock gathered around him, and fol-
lowed closely as he stepped forward carrying his musical bouquet. It is said that for days the sheep listened, and were led on, until the fragile blossoms began to fade and die; then, as the beauty of each withered and passed off, its little fairy chime of sweetness became more subdued and delicate, until the faintest whisper of a sound had softly died away, and never again was heard.

' The azure harebell, that did sweetly ring
  Her wildering chimes to fragrant butterflies.'

We have ten wild species of bell-flowers, but none are more lovely than this our common harebell, or the still more delicate and lighter blue ivy-leaved bell-flower (*Campanula hederacea*), which grows in moist, shady places, but is not gene-

Milkwort—*Polygala vulgaris.*

rally common.

We will now turn down from our moors, and pass through the wood, though the blossoms there are few in comparison to what we gathered in earlier months.

Spreading with creeping root through the mossy

ground, we may pluck the lesser winter green (*Pyrola minor*), its flower-stalk rising some six inches from out its cluster of round shining leaves, and bearing pretty little pale pink blossoms. The white flowers of the elder-tree are now giving place to the berries, and the pretty rose-like blossoms of the bramble have burst into full beauty on its long trailing sprays :

> ' Thy fruit full well the schoolboy knows,
>   Wild bramble of the brake !
> So put thou forth thy small white rose ;
>   I love it for his sake.
>
> Though woodbines flaunt, and roses glow,
>   O'er all the fragrant bowers,
> Thou need'st not be ashamed to show
>   Thy satin-threaded flowers.'

Pluck some of the white flowers of the dog-wood as we go along, and then let us away to the fields ; for the broad-leaved garlic, or ramsons (*Allium ursinum*), makes itself known to other sense than smell. Its clusters of white flowers are pretty, its leaf handsome, and might be taken for that of the lily-of-the-valley, until crushed, and then its disagreeable odour fairly drives us from the woods. There are several other species, usually bearing purple flowers, but none so common as this.

Wandering onwards, we shall likely find a few sprays of the perforated St. John's wort (*Hypericum perforatum*), a bright yellow flower, tipped with small black dots. It has a strong rosiny smell, and owes its name to the old custom of being gathered to make bonfires on the eve of St. John. It is useful in domestic surgery, and in olden times was considered an antidote to all witcheries, and a protection against accidents from thunder and lightning. Many legends and superstitions are connected with this plant :

> ' St. John's wort scaring from the midnight heath
> The witch and goblin with its spicy breath.'

Its specific name refers to the minute holes in the leaves, and therefore it has been called

> ' The herb of war,
> Pierced through with wounds, and seamed with many a scar.'

In the same dry pasture or hedge-side blossoms the small, upright St. John's wort (*Hypericum pulchrum*), by far the most elegant of all the species, a tall plant, with flowers smaller than the perforated kind, but equally pretty, with its delicate green leaves, and bright, yellow, clustering blossoms, often tipped and streaked with deep red. There are eleven species, I believe, all yellow flowers, greatly resembling one another, and many with crowded blossoms ; as Cowper tells us,

> ' Hypericum all bloom, so thick a swarm
> Of flowers, like flies clothing her slender rods,
> That scarce a leaf appears.'

Very pretty is the yellow bed-straw (*Galium verum*),

with its light tufts of clustering golden blossoms, and numerous circles of verdant, slender leaves; the wood sage (*Teucrium scorodonia*), with its spikes of greenish-brown flowers and wrinkled leaves, so much resembling the sage of our gardens ; and the common red centaury (*Erythræa centaurium*),

> 'The crimson darnel-flower,'

a low plant, with clustering red blossoms, and light - green, bitter - tasting stems and leaves.

One of our prettiest vetches is now in bloom, winding its tangled stalks, and clasping tendrils through every hedgeway. This is the tufted vetch (*Vicia cracca*), with its rich purple-blue flowers and long silky leaves. There also is the kidney vetch, or lady's fingers (*Anthyllis vulneraria*), which we gathered in May, with its crowded heads of yellow flowers growing in pairs at the end of each stem, from the soft woolly flower-cup ; as well as the meadow vetchling (*Lathyrus pratensis*), which frequents the moist meadow, making it gay with its bright yellow blossoms. Clare says :

YELLOW BED-STRAW—
Galium verum.

'The yellow vetchling I have often got,
  Sweep creeping o'er the banks in sunny time.'

By far the loveliest of all our vetch tribe, and certainly
one of our most elegant climbers, is the wood-vetch
(*Vicia sylvatica*), which weaves its long stems and
branched tendrils through the tangled wood, often
climbing to the height of six feet. Its flowers are a
delicate white, daintily veined with bluish lines; and
most lavishly does it spread its blossoms during this
and the next month:

'Where profuse the wood-vetch clings
  Round ash and elm in pencilled rings,
  Its pale and azure-pencill'd flower
  Should canopy Titania's bower.'

The hedge woundwort (*Stachys sylvatica*) is also now
in flower, raising its spike of purple blossoms streaked
with white, and growing in whorls around a stem some
three feet high. Its leaves are like the stinging nettle,
but soft and downy. It has an unpleasant odour, and
only the snail is known to feed upon it. There are
several species of woundwort blooming in this month,
and resembling one another very closely.

Another disagreeable-smelling flower we may now
gather in the common black horehound (*Ballota nigra*).
It is a very common plant, rising often four feet high;
has dull red blossoms growing in whorls, and dusty,
grey-green leaves.

Here is the Deptford pink (*Dianthus armeria*), its

rose-coloured flowers dotted with white, growing in bunches on long stiff stems ; the great hairy willow-herb (*Epilobium hirsutum*), a tall, handsome plant, with soft, downy leaves, rich purple blossoms,

'And fruit-like scent so mellow.'

The self-heal (*Prunella vulgaris*), with its heads of bluish-

PHEASANT'S-EYE—*Adonis autumnalis.*

purple flowers and purple-tinged bracts growing amongst its blossoms ; the common cow-wheat (*Melampyrum pratense*), which rises about a foot high, has slender straggling branches, and pale, sulphur-coloured flowers ; the pinkish lilac blossoms of the common calamint (*Calamintha vulgaris*), so often made into tea by villagers, and whose Greek name signifies *good mint;* and the beautiful crimson pheasant's-eye (*Adonis autumnalis*), its rich buttercup-shaped flower, with its handsome purple centre, gleaming brightly from amongst its many narrow, branched green leaves, so cleft into segments. An old botanist tells us it was sold by the herb women in the London markets under the name of *rose-a-rubie;* and we often hear it named *flos Adonis*, and Adonis flower, in consequence of the old fable, that it sprang from the drops of blood which fell from the unfortunate Adonis, who was killed while hunting a wild boar, and of which I told you when we gathered the wood-anemone. In remembrance of this legend, it is to this day familiarly called by the French, *gouttes de sang*—drops of blood.

It often adorns the corn-field, where the brilliant blue succory, or chicory (*Cichorium Intybus*), will sometimes bloom, and prove a noxious weed to the farmer. It is a handsome flower, rich in its deep blue tint, and growing often three feet high.

> ' The chicory flower, like a blue cockade,
>   For a fairy knight befitting.'

A variety of this plant is extensively cultivated in France and 'Germany, its carrot-like root being roasted, and used in the adulteration of coffee.

L

We must not omit to gather the premorse scabious, or

CORN BLUE-BOTTLE—
Centaurea cyanus.

devil's-bit scabious (*Scabiosa succisa*), which blooms in this month, and is common on meadow or pasture. Its name is derived from *scabies*—leprosy. Some species have been considered a cure for that disease. It grows about a foot high. Its round heads of flowers are a deep bluish purple, its leaves toothed, its root fleshy, and as though broken or bitten off at the end, and from this has originated its name. Gerard tells us : 'Old fantasticke charmers report that the divel did bite it for envie, because it is an herbe that hath so many good vertues, and is so beneficial to mankinde.'

Amongst the corn that waves in the summer breeze, fast ripening 'neath the summer sun, and adding such beauty to the landscape, we cannot fail to notice the beautiful corn blue-bottle (*Centaurea cyanus*), that raises its erect head among the ears of wheat, contrasting its exquisite tints of blue with the less showy colouring around.

> ' No summer sky hath a more delicate hue
>   Than thy blossoms, that ope 'mid the rip'ning corn,
>   Nor the veins, the clear sky shining through,
>   Of the fairest maiden that e'er was born.'

It usually grows from two to three feet high, and is as elegant in form as it is brilliant in hue, with its pretty scaly calyx, its outer florets ranged so exquisitely around the smaller ones within the disk, and its stems and green leaves so thickly covered with a cottony down. A beautiful blue colour may be expressed from the plant; but as its tint is not permanent, it is not now used in dyeing. In Scotland it is termed blue-bonnet, in France *bluet;* and in former days we called it hurt-sickle, because it was supposed to blunt the edge of the reaper's hook. Its botanical name is *Centaurea cyanus.* It has also been named *Chironium;* both appellations being derived from the Centaur Chiron, who, happening to be wounded in the foot by the fall of an arrow when he was entertaining Hercules, applied the plant to the sore, which was speedily healed.

> ' The centaury   .   .   .   .
>   That from Thessalian Chiron takes its name.'

Now my young friends may not know what was a Centaur, or who was Chiron, and therefore will feel interested to hear that the Centaurs were an ancient people of Thessaly, who tamed their wild horses, and rode about upon them, often hunting wild bulls, or gathering in their stray cattle. In consequence, they frequently were seen on the borders of surrounding countries; and the sight was so strange to their ignorant neighbours, that they believed them to be monsters, half man and half beast. Chiron was one of these Centaurs, famous alike for his wonderful knowledge in astronomy, music, and the polite arts. He is said to have taught mankind the medicinal use of plants. He was the preceptor of Hercules, whom he rendered the most valiant and accomplished hero of the age. Chiron was greatly beloved by Hercules, who afterwards had the misfortune, when engaged in battle with the Centaurs, to wound with a poisoned arrow his old preceptor in the knee. How it was that our beautiful corn-field blossom was not again applied, or whether it had lost its former efficacy, I do not pretend to say; but we are told that, after excessive agony, Chiron died, and Hercules, frenzied with grief and horror, slew every Centaur present, nor rested till he had utterly destroyed the race from the face of the earth.

On the waste ground we may gather the small lilac blossoms of the vervain (*Verbena officinalis*), a tall, slender plant, with few leaves, called in olden times the holy herb.

> ' Bring your garlands, and with reverence place
> The vervain on the altar.'

The Druids held this plant in almost as high veneration as the mistletoe. It was used to purify their altars; and Mason describes the solemn incantation:

> ' Lift your boughs of vervain blue,
> Dipped in cold September dew,
> And dash the moisture chaste and clear
> O'er the ground, and through the air :
> Now the place is purged and pure.'

It was held a sacred plant among the Romans, and worn by the ambassadors when declaring war.

> ' A wreath of vervain heralds wear,
> Amongst our gardens named,
> Being sent that dreadful news to bear,
> Offensive war proclaimed.'

Near it we may find the poisonous henbane (*Hyoscyamus niger*), so often smoked by country people as a cure for toothache. Its dingy, cream-coloured flowers, growing in clusters, are beautifully veined with purple, like delicate network spreading over the petals, that sit close to the footstalks of the leaves, which are downy and viscid. It is a powerful narcotic; its smell is even said to be dangerous to some:

> ' A poisonous mischief it is thine to work ;
> Baneful the juices that within thee lurk.'

Nearly all animals avoid it, excepting swine, which are said to relish it; and from this it is supposed to have derived its botanical name, which is taken from the

Greek, and signifies *pig's-bean*.   Bishop Mant well describes it and the deadly nightshade:

> 'Laced with purple veins,
> Fair to the sight, but by the smell
> Unprized, the henbane's straw-tinged bell,
> With danger fraught.   But e'en more full
> Of danger, dark of hue, and dull
> Of aspect, near, with purple flowers,
> Perchance the deadly nightshade lowers.'

This is the *Atropa belladonna*, or deadly nightshade, of which we gossipped last month.

> 'Growing on the weedy bank,
>   We shall find the hemlock rank'

(*Conium maculatum*), easily recognised by its disagreeable odour, its tall, smooth stems spotted with brown, and bearing umbels of numerous white flowers.   It sometimes grows as high as six feet.

Many thistles are now in bloom—too many, I fear, for us to observe all, though their flowers are handsome, mostly blue or yellow.   They do not all belong to one genus, and are difficult to distinguish, except by botanists. In many places the young leaves are eaten as spinach. There is the tall marsh sow-thistle (*Sonchus palustris*), with its large yellow blossoms and hairy flower-cup; the fragrant musk-thistle (*Carduus nutans*), with its rich purple-red flower, which droops from its stem of three feet; the milk-thistle (*Carduus marianus*), a very prickly plant, with deep purple flowers, and distinguished by the milky-white vein that intersects its dark spiny leaf; and the star-thistle (*Centaurea calcitrapa*), which, however, is scarcely a true thistle, for its sharp spines deck its flower-cup only. Every schoolboy is familiar with the common sow-thistle (*Sonchus oleraceus*), whose milky juice is so relished by the tame rabbit. The carline thistle (*Carlina vulgaris*) may be distinguished by its yellow flowers, with their inner circlet of purple leaves; the real cotton thistle (*Onopordum acanthium*) by its white woolly covering. This last is a majestic plant, frequently cultivated in gardens, and often mistaken for the Scotch or common thistle (*Cnicus lanceolatus*), the national insignia of Scotland, so abundant in that country. Well does it merit the defiant motto affixed to it: '*Nemo me impune lacessit*'—'No one touches me with impunity;' or, as we are told the Scotch interpret the Latin, 'Ye mauna meddle wi' me.'

> ' I am Sir Thistle, the surly,
>   The rough, and the rude, and the burly ;
>   I doubt if you'll find
>   My touch quite to your mind,
>   Whether late be your visit or early.'

The down on the thistle stems and leaves are gathered by villagers for tinder, and the seeds much loved by birds.

In the corn-fields we may gather the handsome blossoms of the field knautia (*Knautia arvensis*), which has dark green, hairy leaves, and large heads of purplish lilac flowers, growing on a stalk two or three feet high. It is said these blossoms will change to a pretty green colour if held in tobacco smoke. We have but this one species.

Here also we may find the red bartsia (*Bartsia odontites*), a common, unattractive flower, of a dull purple colour; as well as the dyer's green-weed (*Genista tinctoria*), a tall, light yellow, butterfly-shaped blossom, yielding a good dye, its scientific name being derived from the Celtic *gen*—a small bush; and the spotted persicaria (*Polygonum persicaria*), with its pretty spikes of small reddish-pink flowers, and long slender leaves, bearing a dark purple spot on each, which, tradition would have us believe, was caused by a drop of blood falling on a plant which was growing at the foot of the cross. Another species blooms just now, called the amphibious persicaria (*Polygonum amphibium*); and very beautiful are its spikes of rose-coloured blossoms, blooming by the streamlet's side. The genus has derived its botanical name from the knotted joints in the stems,— *polygonum* being taken from the Greek, and signifies *many joints*.

Along by the sea-coast we might gather many flowers; but we have done enough for this day's ramble, and perhaps we may hope, some future time, to enjoy a wander on the shore. July blossoms are so abundant, we cannot gather half her store of 'buds and bells,' scattered in rich profusion on every side.

'Springing in valleys green and low,
   And on the mountain high,
And in the silent wilderness,
   Where no man passeth by.

Our outward lives require them not;
   Then wherefore had they birth?—
To minister delight to man,
   To beautify the earth;

To comfort man, to whisper hope,
   Whene'er his faith is dim;
For whoso careth for the flowers
   Will care much more for him.'

VIII.

AUGUST.

'Brightly, sweet summer, brightly
 Thine hours are floating by,
To the joyous birds of the woodland boughs,
 The rangers of the sky.
And brightly in the forest,
 To the wild deer wandering free ;
And brightly, amidst the garden flowers,
 To the happy, murmuring bee.'

RIGHTLY, indeed, and swiftly, are our beautiful summer months gliding away; and so imperceptibly do the seasons flow into one another, that it is impossible to mark the time of change. In this warm month of August, we may observe that the forest trees are now being touched with their various tints of yellow and brown :

'Autumn laying, here and there,
 A fiery finger on the leaves.'

A ruddier tint has gathered on the orchard fruits ; a

warmer glow is spreading over the swelling grain, bending beneath the soft breeze that travels in long, sweeping waves across the fields; and a rich promise of plenty rests upon the earth.

The woodlands are still decked with the 'gorgeous flowerets' of July. Feathery mosses and luxuriant ferns are spreading over our pathway; but we must not linger over these. Though August does not bring us so many fresh blossoms as the preceding months, yet enough of fair flowers, conspicuous in their golden brightness, have sprung up to form our monthly bouquet:

'Dewy meadows enamelled in gold and in green,
　With kingcups and daisies, that all the year please;
　Sprays, petals, and leaflets that nod in the breeze,
　With carpets, and garlands, and wreaths deck the way.'

We will first gather from the hedges, now decked with the golden cones of the elegant hop (*Humulus lupulus*), which trails its prickly stems in graceful twinings, and droops its fragrant catkins in loose-hanging clusters. Its English name of hop has been derived from the Saxon, *hoppan*, to climb; its generic name comes from *humus*, rich soil, which is needful to its culture. It is valuable as a tonic medicine.

'Awhile regarded as a noxious weed,
　The hop, with tonic properties imbued,
　Was scorned unjustly; surely we had need
　Be careful to condemn, though vile indeed
　May seem the object, through false mediums viewed.

COMMON HOP—*Humulus lupulus.*
173

The hop is largely used in the bitter flavouring of beer, and extensively cultivated for this purpose in many of our English counties, where the luxuriance and elegance of the hop grounds equal in grace and beauty the famous vineyards of more southern climes.

> ' Here swelling peas on leafy stalks are seen ;
> Mixed flowers of red and purple shine between.'

The wild everlasting pea (*Lathyrus sylvestris*) is not a common flower, but very pretty, creeping among bushes, or climbing some feet high—

> ' With taper fingers catching at all things,
> To bind them all about with tiny rings.'

The stems are strong, and each leaf-stalk bears two leaflets and tendrils. Its blossoms grow in clusters, are pink and purple, tinged with green, and look like butterflies turned to flowers.

> ' Thus, year by year, in clusters full of grace,
> Its blooms expand, and beautify the place ;
> Thus, like perennial pleasures, year by year,
> Its clasping tendrils twine and flourish here.'

The only other species of wild pea is that which we mentioned in June (*Lathyrus pisiformis*), and which still creeps over the sandy links of the sea-shore, covering many a barren spot with its pretty papilionaceous blossoms ; for

> ' Nature pencils butterflies on flowers.'

But perhaps the greatest ornament of our hedgerow

is the elegant great bindweed (*Calystegia sepium*), figured on page 126, and described in June, as blossoming during autumn :

> ' The creeping plant, that steals, and steals along,
> And everywhere insinuates itself,'

with its large, white, vase-like blossoms hanging amidst the boughs of the surrounding bushes, and adorning them with the pure beauty of its snowy bells.

Here also we may find the common tansy (*Tanacetum vulgare*), with its waving, dark-green leaves, and yellow clusters of thick, button-like flowers. It is a useful plant in medicine, and strongly aromatic :

> ' The fragrant tansy breathing from the meadow.'

Its young leaves are often mixed in omelets and puddings to form a Lent dish ; but the flavour is strong and unpalatable. It is a tall plant. Clare, in his description of a cottage garden, speaks of the

> ' Tansy running high,
> That o'er the pale-top smiled on passer-by.'

Beside it often grows the wood betony (*Betonica officinalis*), which is also a tall plant, its flowers growing in whorls up the stem. It has few leaves, deep, purple-red blossoms, and somewhat resembles the dead nettle. Country people gather the leaves to smoke as tobacco, or powder for snuff.

The yellow toad-flax (*Linaria vulgaris*), with flowers shaped like the snap-dragon, is now conspicuous,—a gay but troublesome weed, which children call 'butter-and-eggs.' Its expressed juice, mixed with milk, is often used to attract and poison flies.

But perhaps the handsomest blossom of the wayside is the greater knapweed (*Centaurea scabiosa*), which grows tall, with large, bright purple blossoms, and hairy leaves. It is very common; as is also the black knap-weed (*Centaurea nigra*), a smaller and less bright flower; but, as occasional blossoms are to be found as late as Christmas, we shall only gather them in our November ramble.

The wild carrot (*Daucus carota*) may also be seen here, shooting its tall, tough stem upwards, surmounted by its large cluster of greenish-white blossoms, faintly tinged with pink, and decked with its graceful feathery leaves. The root smells exactly like our cultivated carrot, but is a pale yellow-white.

Common, by every roadside, is the common soap-wort (*Saponaria officinalis*), rising a foot high, with large leaves, and pale, rose-coloured flowers, growing in clus-ters. It makes a lather when placed in hot water, and from this is named.

Many species of campanula are conspicuous during this month, blooming in the hedge-bank, or woody place; a genus which, from its bell-shaped flower, has been named *campanula*, a little bell. One of the com-monest, at least in England, is the nettle-leaved bell-flower (*Campanula trachelium*), a handsome, tall, hairy

M

plant, with drooping purple flowers, angular stems, and leaves resembling the nettle. We have ten different kinds of campanula, the delicate harebell (*Campanula rotundifolia*), which we gathered last month, being one of them. The smallest species is the ivy-leaved bell-flower (*Campanula hederacea*), an elegant little plant, often abundant by the side of a stream, with small blue blossoms, bright-green, ivy-shaped leaves, and delicate stems. But the finest and most beautiful of them all is the giant bell-flower (*Campanula latifolia*), towering, in all the beauty of its brilliant blue blossoms, high above the rest,—one of the most showy and abundant flowers of the woody glens of Scotland, though somewhat rare in England, especially in the south. It is often called the giant throat-wort, and by that name is mentioned in *Rokeby:*

> ' He laid him down,
> Where purple heath profusely strown,
> And throat-wort, with its azure bell,
> And moss and thyme his cushion swell.'

It is the Canterbury bell cultivated in our gardens, and its tall spikes of showy flowers often vary from the rich deep purple to a delicate white:

> ' Campanula ! who knows not well ?
> Beauteous Canterbury bell.'

Here, by the hedge-side, is the pale yellow spike of wild mignonette (*Reseda lutea*), which yields such a useful dye, and rears its head upwards of a foot high.

Linnæus affirmed that its blossoms followed the course
of the sun ; looking to the east at sunrise, to the south
at noon ; in the afternoon west, and at night north.
Our wild mignonette is scentless, unlike the fragrant
species of our flower-garden, whose sweetness has won
for it the affectionate name of ' little darling.' Cowper
speaks of

> ' Window-sashes fronted with a range
> Of orange, myrtle, or that fragrant weed,
> The Frenchman's darling.'

But we must not linger over hedge beauties longer.
' Look on the fields, for they are white already to
harvest ;' and many a gay blossom gleams amongst
the waving corn, which is now ripening ready for the
reaper's sickle. A more beautiful sight can scarce be
witnessed. The crimson poppy scatters its falling leaves
over the chaffy-coated wheat ; the little scarlet pimpernel raises its glad face to the genial sunshine. A
stray ox-eye, with its golden centre, and circle of white
leaflets, droops its head among the tall grass in the
hedge-side ; and the showy, starry blossoms of the
corn-marigold (*Chrysanthemum segetum*) are spread in
profusion, making the fields look gay, with their numerous golden flowers. In olden times, they were called
golds or goolds, which we are told means sun-flower ;
and by this name they are often spoken of by the poets,
who also frequently allude to the daily closing of this
blossom. Linnæus says it opens from nine in the

CORN-MARIGOLD—*Chrysanthemum segetum.*

morning until three in the afternoon ; and very beautifully does Shakespeare, in his *Winter's Tale*, tell of

'The marigold, that goes to bed
    with the sun,
And with him rises weeping.'

And again, in *Cymbeline :*

'The winking mary-buds begin
    To ope their golden eyes ;'

whilst Chatterton speaks of

'The mary-budde that shutteth
    with the light.'

The marigold, infused in vinegar, was said to prevent infection ; and the flower alone to remove effectually the pain caused by the sting of a bee. It is a very handsome flower, and one of the largest golden blossoms we have, flowering very plentifully, until checked by winter frosts.

> ' Open afresh your round of starry folds,
>   Ye ardent marigolds !
> Dry up the moisture of your golden lids ;
>   For great Apollo bids
> That in these days your praises shall be sung.'

It grows about a foot high ; and though its brilliant colour is pleasing to *our* eyes, the farmer loves it not, for to him it proves a troublesome weed, which has been known to destroy the crop of a field.

And with little more favour does he look upon the handsome field knautia (*Knautia arvensis*), that raises its large lilac heads above the brown corn, and which we gathered last month.

Another annoying plant in the corn-land is the spurrey (*Spergula arvensis*), a pretty little blossom, with small clustering white flowers and slender leaves.

Late as it is, we may still find, lingering in the shady place, a late blossom of the wood loosestrife, or yellow pimpernel, which we gathered in May, trailing its weak stems and delicate green leaves over the ground.

We have not time to visit the moors to-day, where the flowers of last month still bloom so fresh, and fair, and beautiful,—the brilliant broom, with its butterfly-shaped blossoms, clothing the hill-side as with a gorgeous garment of glittering gold, and filling the soft morning air with its sweet balmy fragrance.

> ' On the low mountain side the purple heath
>   Blossometh freely, swayed by heaven's breath ;

In the low valley, and the rocky glen,
And solitudes afar from haunts of men.'

Thyme and graceful harebell are there ; the little golden-hued, common tormentil (*Tormentilla officinalis*), with its pretty blossom of four petals, rearing its slender stems among the grass, or creeping along the ground ; and that little parasite, the lesser dodder (*Cuscuta epithymum*), twining its long, leafless red threads through the foliage of every bush and shrub, hanging their branches with its clusters of tiny pink-white blossoms.

' The parasite cascula liveth not
By nourishment that its own root supplies,
But, meanly clinging to another,
Absorbs its vital juice.'

Many of the cudweeds (*Gnaphalium*) are now in bloom ; better known, perhaps, to my young friends under the more general name of everlasting. They are silvery-looking flowers, with downy white leaves and stems, and are but rare wildlings in England. The common cudweed (*Gnaphalium germanicum*) is the most frequent, and grows in gravelly pits, or waste stony places. It may readily be recognised by its strange manner of growth. Its stem is about a foot high, terminated by a head of flowers, from beneath which spring several branches, each bearing a bunch of blossoms, these again being divided, and rising higher than the main stem. The handsomest kind is the *Gnaphalium orientale*, a shrubby everlasting, which grows to the height of three feet, and is a soft yellow colour. It is used in Portugal to deck the churches in the winter season ; and in France is dyed a deeper yellow, and black, and formed into wreaths, to adorn the tombs of departed dear ones—an emblem of immortality :

> ' The everlasting flower, wherewith the hand
>   Of fond affection decorates the grave,
>   When in the place of death it takes its stand,
>   Weeping for those it had no power to save.'

Let us now wander down to the little streamlet, that murmurs softly as it steals its way 'neath the overhanging branches, the shrubs, and rushes, tangled with many blossoms that fringe its banks. The flag-flower, the beautiful lily—

> ' The stately flower, the spotless river queen,
>   With silver chalices on stems of green,
>   Uplifted aye to catch the morning dews ;'

the water-violet, the plantain, etc., are now all passing away ; but a few sprays of the lovely, bright-eyed forget-me-not are still lingering for us to gather, nodding their tufted heads to every breath of wind.

Looking down into the water, we discern masses of various grasses, tiny flowers, tufts of silky-jointed threads, through which the bright-scaled fishes and many aquatic insects sport and play, whilst the pond-weeds (*Potamogeton*), with their brown-looking leaves, and the duck-weeds (*Lemna*), with their bright green foliage, float in dense masses on the surface of the stream. And see how, thickly fringing its margin as far as the eye can look, the spiked purple loosestrife (*Lythrum salicaria*) is rearing its tall, richly-coloured blossoms from amongst the rushes, crowding its square stem with whorls of many flowers. Its leaves are a fine rich green ; and altogether it is one of the handsomest of our water-side plants, its tall spikes of purple blossoms bending to every passing breeze. I have heard that these, and not the orchis, were the 'long purples' which formed part of poor Ophelia's garland. The hyssop-leaved loosestrife (*Lythrum hyssopifolium*) also blooms in this month, but is less common— a smaller and not attractive plant, with dark purple flowers.

There we may gather the common fleabane (*Pulicaria dysenterica*), so named from the belief that its smoke is offensive to insects. Its flower is like a yellow star, its leaves downy and pale in tint ; the whole plant about a foot high.

Another common water-side plant is the tall hemp agrimony (*Eupatorium cannabinum*), a large, tall, dull-looking plant, with fragrant flesh-coloured, or purple-pink blossoms growing in clusters, often three feet high, its downy leaves studding thickly the reddish stem. There is but one species.

There is also the Alpine enchanter's nightshade (*Circæa Alpina*), which greatly resembles the common kind we gathered in May, though a smaller plant, with longer stems.

One of the handsomest reed plants is now in flower, for which we shall not long have to look, so common is it by every stream. This is the great reed-mace, or cat's-tail (*Typha latifolia*), which sometimes grows six or seven feet high. Its leaves are a grey-green, its catkins a greenish-brown. In some countries its downy seeds have been used instead of feathers to stuff beds; and mats and baskets are often made of its leaves. This is the plant which Rubens represents as the reed or sceptre which was placed in the hand of our Saviour by the mocking soldiery. The reed was often cut and made into a musical pipe by our shepherds of old,— first invented, as the legend tells us, by Pan, the god of the shepherds:

> ' The lovely Syrinx, when pursued by Pan,
>   Into the river for protection ran;
>   She plunged beneath the wave, and, in her place,
>   Tall, shapely reeds sprang up, the spot to grace;
>   The sylvan god then cut, together bound,
>   And made an instrument of sweetest sound.'

The other species, the lesser reed-mace (*Typha angustifolia*), is smaller and less common. It rarely stands higher than three feet; and very dingy is its fussy-looking club of a blossom.

The sweetest flower of the moist and boggy ground is the common grass of Parnassus (*Parnassia palustris*), with its soft, cream-coloured blossoms, so delicately veined, rising from amidst its long-stalked, heart-shaped leaves. It is a beautiful plant, chaste and elegant, and cannot fail to attract our admiration, its white flowers glancing out brightly, in beautiful contrast with its surrounding leaves.

GRASS OF PARNASSUS—
Parnassia palustris.

> ' Thy simple flower, so finely veined,
>   So delicately marked and stained ;
>   Above man's ingenuity
>   The skill divine displayed in thee.'

There is but one species.

The yellow, or bog asphodel (*Narthecium ossifragum*), blossoms still on boggy or moist ground ; but is very rare, except in Lancashire, where its spikes of yellow flowers may occasionally be gathered. The poet has told us of

> ' Eden's radiant fields of asphodel.'

And again, in Pope's *Ode on St. Cecilia's Day*, we read of the

> 'Yellow meads of asphodel.'

It was the true asphodel—'the flower of the tomb'—which the Greeks used in the decorations of their graves, the ancients believing that the grains which it produces afforded nourishment to the dead.

And now, my dear young friends, our August bouquet is made up; and, ere we enjoy another ramble, summer flowers will entirely have passed away, and the few yellow blossoms of autumn alone be left to greet us. A shade of sadness falls as we bid our good-bye to summer, —to glorious, sunny summer,—with its bevies of glad, happy birds, its rich profusion of varied flowers, its full-leaved, wide-spreading trees.

> ' The summer time ! the summer time !
>   The noontide of the year ;
> Oh, can there be a human heart
>   To which it is not dear ?
>
> Then blessings on the summer time,
>   Its sunshine and its flowers ;
> I love its wide, brown moorland wastes,
>   And its shadowy greenwood bowers ! '

Oh, how bright, and true, and pure is the enjoyment of our summer hours, when, luxuriating in the beauties of nature, we read the great volume of God's handi-work, which tells us,

> ' There is a tongue in every leaf,
>   A voice in every rill ! '

We look on 'the grass of the field,' of which Jesus hath said, ' If God so clothe the grass of the field, shall He not much more clothe you, O ye of little faith ?' We ' consider the lilies,' and all the gorgeous array of summer flowers that are scattered without stint ' o'er hill and

dale,' rich in their fragrant beauty, raising their glad
faces to the sky, and each one, down

> ' To the green blade that twinkles in the sun,
> Prompts with remembrance of a present God.'

From contemplating 'the stars of earth,' each one of
which seems to breathe of heavenly things, our minds
naturally soar to 'high and heavenly places,'

> ' Where ever-verdant spring abides,
> And never-fading flowers ; '

and our hearts yearn for that home of many mansions
which Christ is preparing for us :

> ' Jerusalem, the golden,
> Methinks each flower that blows,
> And every bird a-singing,
> Of thee some secret knows.
> My thoughts, like palms in exile,
> Climb up to look and pray
> For a glimpse of that dear country
> That lies so far away.'

IX.

SEPTEMBER.

———◆———

'Spake full well, in language quaint and olden,
  One who dwelleth by the castled Rhine,
When he called the flowers so bright and golden,
  Stars that in earth's firmament do shine.

Wondrous truths, and manifold as wondrous,
  God hath written in those stars above;
But not less in the bright flow'rets under us,
  Stands the revelation of his love.'

OETHE it was who called the flowers
' stars of earth,' and it is to him that
Longfellow refers in the lines above.
Surely at no season is the beautiful
simile more applicable than during
autumn, and especially this month of
September, when the blossoms are mostly
yellow, from the beautiful broom, the pride
of the otherwise dreary heath, to the few stray

'Buttercups that *will* be seen,
  Whether we will see or no.'

September, with its pure, genial air, and soft, warm sunshine, is a delightful month, though it brings but few flowers to supply the places of our previous months' profusion ; and we look with sadness on the fast-fading summer blossoms, feeling we have reached

> ' The season when the green delight
>   Of leafy luxury begins to fade,
>   And leaves are drooping hourly on the sight.'

The shady summer is gliding imperceptibly into the gorgeous colours of autumn ; fruit and berries are taking the place of flower and leaf :

> ' The autumn sheaves are on the hill,
>   And solemn are the woods, and still,
>   With clustering nuts on every bough.'

Ferns are thickening in the woods, nestling in every nook and crevice, waving their luxuriant leaves, in all their feathery grace and beauty, around the perishing trunk of some blighted tree,—

> ' A halo hovering round decay.'

The mosses which carpet the ground are a study of themselves, so soft, and fresh, and delicate. Fully should we be repaid in devoting a ramble to their exclusive examination ; for

> ' There's beauty all around our path,
>   If but our watchful eyes
>   Can trace it 'midst familiar things,
>   And through their lowly guise.'

The morning dew is now heavy on the grass :

> ' Dewdrops, like diamonds, hang on every tree,
> And sprinkle silvery lustre o'er the lea ;'

where the round buff mushroom, with its rosy-fringed under surface, is gleaming on the meadow :

> ' Up in a night the mushroom springs,
> And who but he must be king of the mead ?
> One loves not the sight of such upstart things ;
> Quickly they rise, and they fall with speed.'

But let the gatherer be careful they are *real* mushrooms which he picks, for many a fatal error has occurred in mistaking poisonous fungi.

> ' So much alike, the wholesome and the bad,
> That with suspicion we must ever look
> Upon all members of the mushroom tribe.'

The various grasses are very beautiful now, in their rich brown ripening perfection ; some draped in elegant feathery beauty, others decked with little nodding bells, which I remember as children we used to call ' dothering dollies.'

Under the name of *grasses* are included all the various kinds of grain—barley, wheat, etc. ; but at present I was referring only to the *fancy* grasses, as I would name them, which adorn the hedgeway in such delicate profusion. Though we have not time to examine each, let us gather a few for our bouquet, to which they will add much grace ; remembering, however, how invaluable are the numerous and more productive grasses to the service of man, constituting not only his own staple food, but 'feeding the cattle upon a thousand hills' in the spring and summer months, and providing for winter an inexhaustible supply of sustenance in the form of hay. 'The eyes of all wait upon Thee, O Lord; and Thou givest them their meat in due season.' 'Because Thy loving-kindness is better than life, my lips shall praise Thee.'

The most singular plant of this month is the meadow saffron (*Colchicum autumnale*), with its delicate lilac, crocus-like blossom, rising on its slender stem from amidst the meadow grass. It is not a common flower, and our only native species. It blooms in autumn; but its fruit and leaves do not appear until the succeeding spring. The seed lies buried in the bulb, protected from winter snows, to rise with spring sunshine; and, ripening during summer, produces its blossoms in autumn. In this a striking evidence, fraught with instruction, is given us of the protective wisdom of our divine Creator.

'The well-ordained laws of Jehovah'

will not suffer that the seed of this late-blooming flower should perish in the nipping frosts, as it inevitably would do if left to follow the usual course of nature. Truly may we exclaim,

'Where'er I turn, whate'er I see,
Reveals some work, O Lord, of Thee.'

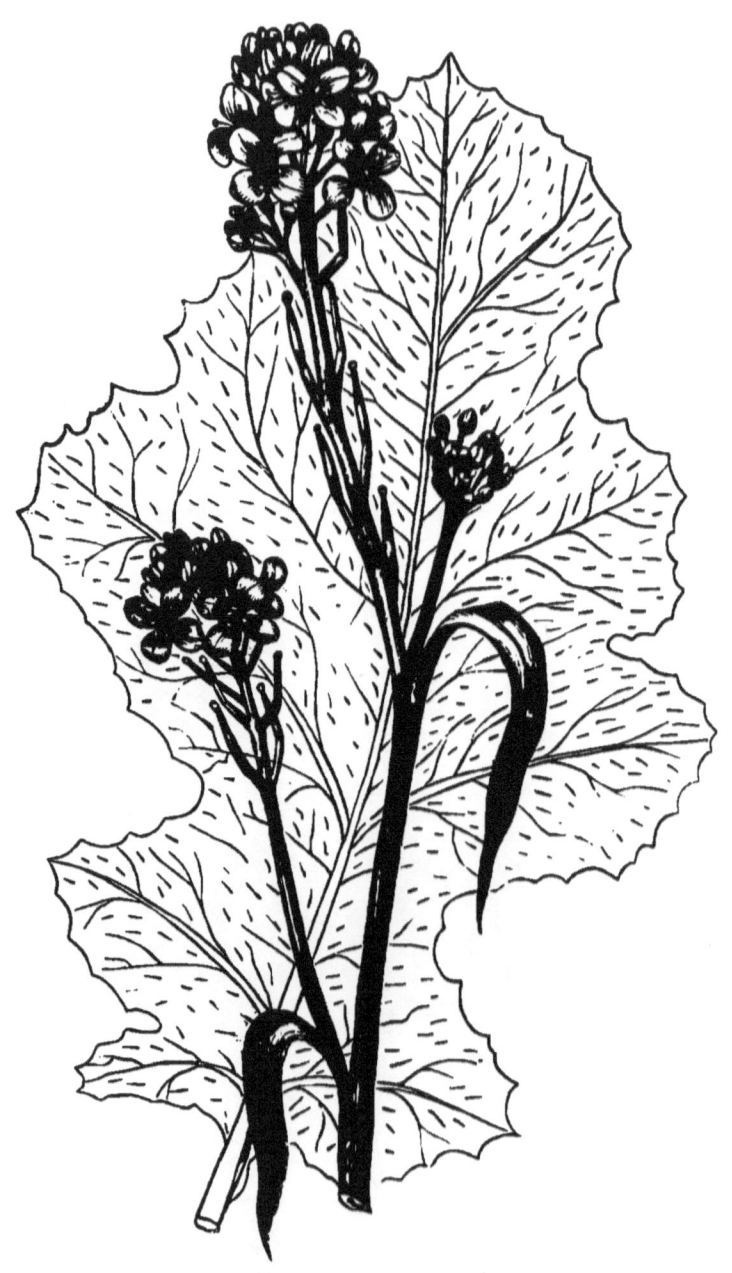

COMMON MUSTARD—*Sinapis nigra.*

194

The meadow saffron possesses strong medicinal virtues,

> ' Qualities which, temperately used,
> Will benefit, but poison if abused.'

It is given in the alleviation of gout and rheumatism. The French call it *morte aux chiens;* but no animal will touch the plant when fresh. It does not appear, however, to be rejected when cut down with the summer grass. The saffron crocus (*Crocus sativum*), from which the drug of that name is obtained, is also now in bloom; but it is a doubtful wildling. It resembles the meadow saffron in colour and manner of growth.

The common mustard (*Sinapis nigra*) grows very abundantly about the borders of fields and on waste ground; and it is well you should recognise so useful a plant. Its stem is upright, with many smooth, spreading branches, decked with small lance-like leaves. The lower leaves are very large, rough, and lobed; the flowers a brilliant yellow, and very numerous. Every one knows what a wholesome stimulant is the mustard, which is made from the seed contained in those little pods you see sitting erect and short, so closely to the stem.

In the pasture or wood now blossoms also, in full perfection, the tall, handsome golden rod (*Solidago virgaurea*). Its bright yellow flowers are crowded in clusters, and butterflies and moths hover around it in vast numbers. It was supposed to possess valuable healing properties, as its botanical name implies, being derived from *solidare*, to unite; and in the days of Queen Elizabeth was sold in London by herb-dealers. But the great glory of the September month is in the harvest, which has now become general; therefore let us hasten to the fields, failing not to gather from the hedgerow as we pass along the country lanes, now being decked with the glitter and brightness of ripening heps and haws.

First we will pluck a spray or two of the knotgrass (*Polygonum aviculare*), which has been blossoming since May—the 'hindering knotgrass,' as Shakespeare calls it, from its rapid, spreading, entangling growth. It has tiny red and white blossoms, sitting so closely on the stem as to be almost hidden by the leaves. It is more common than pretty, and is to be found anywhere and everywhere, even springing up between the paving-stones of the unfrequented streets:

> 'By the lone quiet grave,
> In the wild hedgerow, the knotgrass is seen,
> Down in the rural lane,
> Or on the verdant plain,
> Everywhere humble, and everywhere green.'

It is a valuable little plant to the many birds it nourishes; and cattle, especially sheep, are fond of it. Milton tells us,

> 'The chewing flocks
> Had ta'en their supper of that savoury herb,
> The knotgrass.'

And on this account it is called *grass*, though bearing no resemblance to a true grass. It gains part of its name also from the knottiness of its stems.

Another of the commonest blossoms of our roadside we failed to gather last month, but which is still in bloom, is the common yarrow or milfoil (*Achillea mille-folium*), with its pretty hairy-backed leaves cut into numerous segments. It grows upwards of a foot high, and is called milfoil, from its countless division of leaves, and clustering mass of white or pinkish-purple flowers :

> ' The mille-foil, thousand-leaved, as heretofore,
>   Displays a little world of flow'rets grey.'

There is a rhyme which says,

> ' Achillea millefolium, the name
>   Of one who counted war a glorious game,
>   Was unto thee applied in ancient time ;
>   And hence we liken thee
>   To war, th' epitome
>   Of human folly, misery, and crime.'

It was supposed to possess the quality of stopping any flow of blood ; hence its old name of nose-bleed, or its more common one still of old man's pepper, from the pungency of its leaves.

A larger species, flowering also at this time, is the goose-tongue, or sneeze-wort yarrow (*Achillea ptarmica*), its white blossoms growing often two or three feet high :

> ' If thy dried leaves we pulverise,
>   The sluggish brain to please,

> Thou dost the startled sense surprise,
>   As sneeze quick follows sneeze.'

The woolly milfoil (*Achillea tomentosa*) has bright golden flowers, and is common in Scotland. These comprise our three species of yarrow.

In the hedge-side also blooms the mugwort (*Artemisia vulgaris*), a plant well known for its supposed charm against ague, and valued by country people in cases of consumption ; hence the rhyme :

> ' Why should maidens die,
>   When the nettle grows in March,
>   And the mugwort in July ?'

It is largely employed in Sweden, and also in Ireland and Wales, in the place of hops for flavouring beer, on account of its peculiar bitter taste. We often hear it quoted,—

> ' Bitter as wormwood to the taste.'

It is tall and erect, with purplish flowers, and leaves dark green on the upper surface, but white and cottony underneath. The common wormwood (*Artemisia absinthium*) flowers also at this time, and is common about waste ground. Its leaves are downy, divided into segments, its pale yellow flowers hang in leafy clusters, and the whole plant is usually about a foot high. The common sea-wormwood (*Artemisia maritima*) is also now in blossom by the sea-coast.

Here, too, is the common mallow (*Malva sylvestris*), its large, handsome, purplish blossoms, and round, showy leaves, adorning our lanes with their rich beauty :

> ' The mallow purpling o'er the pleasant sides
> Of pathways green.'

Children call its young fruit cheeses ; and the French infuse its leaves as a pleasant drink or medicine, a cure for cold or cough. The dwarf mallow (*Malva rotundifolia*) also blooms now, and may be recognised by its small, pale, lilac-pink flowers, and roundish leaves ; and the handsome musk mallow (*Malva moschata*), which has large, beautiful, rose-coloured blossoms, divided leaves, and upright stems. The marsh mallow (*Althœa officinalis*) now blossoms by the sea-side, and attracts attention by its handsome flowers and height of two feet. Its thick silky leaves are downy and soft as velvet ; its blossoms a pretty, delicate rose-colour, or purple-blue. It yields, in hot water, an abundance of starchy mucilage, and is said to possess healing properties :

> ' To heal the wound, the grove to decorate,
>   These were its offices in days gone by ;
> What wonder, then, that we should consecrate,
>   The blooming mallow to humanity ?'

The mucilage taken from its root is made into a paste in France, called *Pâte de Guimauve.*

The sea campion (*Silene maritima*) is abundant on the coast during this month ; and very handsome are its tall, solitary, white flowers. Its leaves are small, its root creeping.

The thrift (*Statice armeria*), so familiar as a garden
border, flourishes also on the waste-looking sea-links,
adorning many an elevated crevice or height, and
brightening large patches of dreary land ;

> ' Flourishing so gay, and wildly free,
>   Upon the salt marsh by the roaring sea.'

Its flowers, which are a pinkish lilac, grow in heads, or
tufts, on a downy stalk a few inches high. Its narrow
grassy leaves all spring from the root. Requiring little
nourishment, it thrives anywhere, and thus well merits
its name. Another species is

> ' The sea-lavender, which lacks perfume '

(*Statice binervosa*) ; a sort of everlasting, with pretty
lilac-blue blossoms, and bluish-green leaves.

But this is a digression from our hedge-side and fields,
where we may find one or more of the hemp-nettles.
The large-flowered hemp-nettle (*Galeopsis versicolor*) is
common in the Highlands of Scotland, though somewhat
rare in England. It is a tall, handsome plant, with
bright yellow blossoms, having a purple spot on the
lower lip. The common hemp-nettle (*Galeopsis tetrahit*)
has much smaller flowers, a delicate purplish-yellow or
white, growing in whorls around the stem, which is much
swollen beneath its couplet leaves. The red hemp-nettle
(*Galeopsis ladanum*) is a smaller plant, less common.
Its flowers are a mottled mixture of purple and crimson,

growing in clusters of two or three. Its leaves are small and notched.

On this waste bank we may gather a few sprays of the fool's parsley (*Œthusa cynapium*), a poisonous weed, often mistaken for the true parsley. Its small white petals, and smooth, deep-green leaves, will add a freshness to our bouquet, as well as a sprig of the tall fennel (*Anethum fœniculum*), with its golden blossoms and drooping leaflets :

> ' The fennel, with its yellow flowers,
> That, in an earlier age than ours,
> Was gifted with the wondrous powers,
> Lost vision to restore.'

But lo! here we come upon the beautiful corn-fields, and the reapers' merry voices ring through the sweet September air. Mount this stile, and look upon the abundant, glowing landscape, and say, can there be a fairer sight? The rich brown wheat, with its myriad bulging sheaths of opening wealth, and brown, drooping leaves, ranked in long colonnades, bends and waves in solemn state, murmuring softly with every rustling breeze. The first deep ranks have fallen 'neath the reaper's sickle. There is a busy anxiety to have the harvest quickly gathered in. Here we see 'the binding into bundles,' the piling together of the sheaves, the harvest-cart leading away to the garner, the maidens and young children following after, gleaning as they go, filling their aprons with the stray heads of wheat, reminding one of gentle

Ruth, who gleaned in the field of her kinsman Boaz, none reproaching her. The pale stubble of the field looks bereaved and desolate; for the many beautiful corn-flowers have fallen 'neath the shearer's hook, and now lie faded on the ground, or droop in dimmed beauty amongst the sheaves. Just a few starry blossoms have escaped the ruthless sickle, and gleam bright and golden still,

> ' Cheering, through the shortening day,
> Autumn with her weeds of yellow.'

There is a starry blossom of the bright ragwort (*Senecio Jacobæa*), with its ragged, shabby-looking leaves; the little field madder (*Sherardia arvensis*), its small, pretty lilac blossoms almost hidden; its stiff, sharp, whorled leaves lying close to the ground. You must not suppose this is the dyers' madder (*Rubia tinctorum*), from which so valuable a dye used to be extracted, though our little plant greatly resembles it in appearance. The botanical name *rubia*, taken from *ruber*, signifies *red*. But the cochineal insect is now much more used than the plant in dyeing.

The chicory (*Cichorium intybus*) still shows its cheerful, bright blue blossoms; and there is the dull-looking red bartsia (*Bartsia odontites*), an unattractive plant, some eight inches high, yet possessing a pretty little flower when examined. It is very common in the corn-field, but fades very quickly after being gathered. The yellow bartsia (*Bartsia viscosa*) is a much handsomer flower,

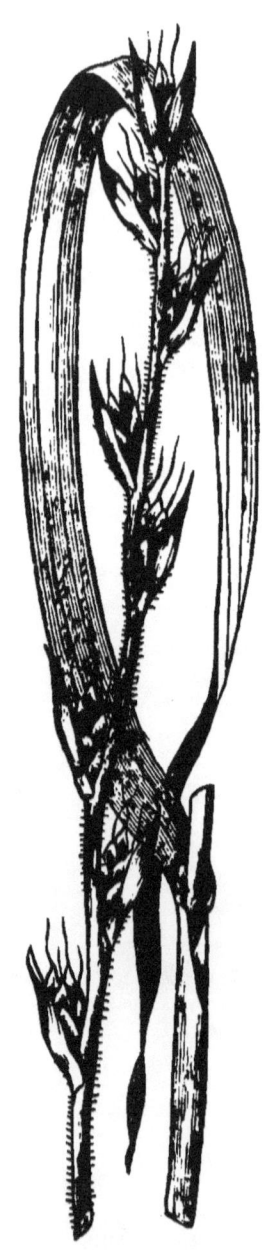

DARNEL—*Lolium temulentum.*
204

but rarer. It has large yellow blossoms stained with purple, round downy stems, and a strong odour of musk. It is plentiful in the west of England.

So we see,

> ' The summer flowers are lingering yet,
> But soon they must decay ;
> Friends of the year, can we forget,
> Or see them pass without regret
> From us away ?'

We must not neglect the pretty May-weed (*Pyrethrum inodorum*), so softly scented, which, in spite of its spring-like name, is an autumnal flower, and often decks the fields as late as December.

Before quitting the corn-fields, my dear young friends, I should like to draw your attention to the darnel (*Lolium temulentum*), which not unfrequently grows amongst the wheat, and which it resembles so closely as often not to be distinguished from it until the ear is formed. It has rather broad leaves, and its thick stem rises as high as the corn. It is supposed to be the tare of Scripture, and possesses peculiarly deleterious properties, rendering it most unwholesome when mingled with flour.

> ' As darnel chokes the rising corn,
> And blights the farmer's hope ;
> So vice, in youthful bosoms born,
> When it for growth hath scope,
> Makes vain the promise of the morn.'

And I am here reminded of a passage in Scripture, over

which my young friends may have wondered, and which they will doubtless remember in Genesis—how Pharaoh dreamed, 'and behold seven ears of corn came up on one stalk.' Now, however extraordinary such an occurrence would seem with us, it was no unusual thing in Egyptian wheat.

And now, as we wend our steps homeward, let us raise our hearts in glad thankfulness to the beneficent Creator, who has granted such wonderful fertility to our earth, and this year blessed us with so rich and abundant a harvest; 'who crownest us with loving-kindness and tender mercies,' and 'hath filled us with the finest of the wheat.'

> ' Lord of the harvest, once again
> We thank Thee for the ripened grain ;
> For crops safe carried, sent to cheer
> Thy servants through another year ;
> For all sweet, holy thoughts supplied
> By seed-time and by harvest tide.
>
> Daily, O Lord, our prayers be said,
> As Thou hast taught, for daily bread ;
> But not alone our bodies feed,—
> Supply our fainting spirits' need.
> Oh, Bread of Life, from day to day
> Be Thou their comfort, food, and stay !'

X.

OCTOBER.

———◆———

'Flowers of the field, how meet ye seem
   Man's frailty to portray !
Blooming so fair in morning's beam,
   Passing at eve away !
Teach this, and oh ! though brief your reign,
Sweet flowers, ye shall not live in vain.'

CTOBER has now arrived, with its shortening days and cooler breezes. The brightness of the year is over, and a sadness has stolen over the landscape. The trees are touched with red and withering brown ; the first yellow leaves are softly stealing down :

     'These fading leaves,
That, with their rich variety of hues,
Make yonder forest in the slanting sun
So beautiful.'

'A fiery finger' rests upon the beech-tree ; soon the ash and poplar—

> ' Each like a fleshless skeleton, shall stretch
>   Its bare brown boughs ;'

and we are reminded of the prophet's words: 'We all do fade as a leaf; and our iniquities, like the wind, have taken us away.' The corn is all cut and safely garnered; and the stubble fields look bare and desolate, save where a delicate harebell raises its head in trembling solitude. The thrush and blackbird are silent in the wood, the skylark has warbled his last farewell, and the twittering swallows homeward fly. The summer flowers are now passing rapidly from us, and we grieve to bid them adieu; but the chill of autumn winds is fast mowing down our few remaining blossoms, and we ask in vain,

> ' Where are the flowers, the fair young flowers, that lately sprung, and stood
>   In brighter light and softer airs, a beauteous sisterhood ?'

Alas ! all are drooping, fading, dying ; yet still is nature beautiful, even in the beginnings of her decay ; for,

> ' Wheeling her flight through the gladsome air,
>   The spirit of beauty is everywhere.'

The hedges gleam with the bright colouring of birds' provisions. The scarlet heps and crimson haws glisten, in contrast with the clear berries of the nightshade, the red fruits of the bryony, the bright orange-coloured seeds of the spindle-tree, and deep red clusters of the berries of the guelder rose.

The orchard boughs are bent beneath their weight of

ripened fruit, now to be gathered and stored away,—
rows of rosy apples and freckled pears; the trees are to
be shaken, and pattering nuts to be picked and stowed in
sacks for winter use; and the black fruit of the bramble
(*Rubus fruticosus*) to be plucked and converted into
conserve and jam for pies and tarts. The hedges are
now full of these trailing brambles, loaded with ripe
blackberries; and we may refresh ourselves with some of
the wholesome fruit. In doing so, who is not reminded
of that sweet pathetic ballad of the *Babes in the Wood*,
whose

> 'Pretty lips with blackberries
> Were all besmeared and dyed;
> And when they saw the darksome night,
> They sat them down and cried?'

The fruit of the bramble is often made into wine, and
in Provence (France) is used to give a deeper colour to
particular wines. The roots, dried and then infused, are
said to form an excellent remedy against an obstinate
cough. I have heard of its branches being used for
cords. So you see, my young friends, the bramble, so
often despised, is a somewhat useful plant, possessing
several advantages. It was the bramble which the trees
of Lebanon chose to rule over them, in that parable spoken
by Jotham to the men of Shechem (Judges ix. 8–15).

The beautiful arbutus, or strawberry-tree (*Arbutus
unedo*), is now in blossom; and nothing can exceed the
richness of its dark, glossy, evergreen leaves, or exquisite

O

delicacy of its greenish-white flowers, which hang like little waxen bells beside the ripening scarlet berries, the produce of last year's blossoms.

> 'The leafy arbute spreads
> A snow of blossoms, and on every bough
> Its vermeil fruitage glitters to the sun.'

Its leaves remain all winter, and in spring are pushed off by the shooting of fresh ones. Thus the branches are ever clothed; and when other trees are losing their beauty, this is in its fullest perfection of loveliness.

> 'Great spring before,
> Greened all the year; and fruits and blossoms blushed
> In social sweetness on the self-same bough.'

This shrub is uncommon as a wildling in England, but is often seen decking the Irish landscape. The fruit resembles the strawberry, but is firmer, and richer in colour. Its specific name, *unedo*—I eat one—is said to have originated in the fruit being too unpleasant for any one to desire a second taste; yet the Spaniards convert it into jam, and the Irish peasantry gather it for sale from the many bushes in the neighbourhood of Killarney.

There are two more species of this berry, the red bear-berry (*Arbutus uva-ursi*), a dwarf shrub, and native of our mountainous heaths, with evergreen leaves, rose-coloured blossoms in short terminal clusters, and smaller berries, a brilliant red, which are much relished by grouse and game; and the black bear-berry (*Arbutus alpina*), common in the north of Scotland, a trailing

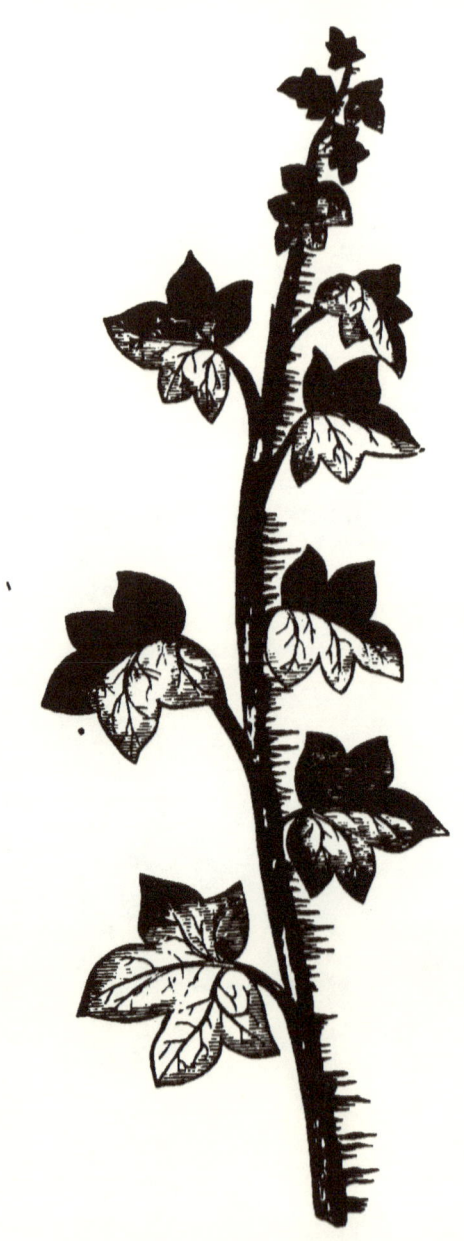

COMMON IVY—*Hedera helix.*

212

shrub, with brilliant leaves, coloured in autumn with a beautiful crimson tinge.

The ivy (*Hedera helix*) is one of our latest flowering plants, and now puts forth its pale green blossoms, over which, on sunny days, the many bees and brilliant-winged butterflies hover in swarms. The ivy, which decks with grace and beauty many a hoary ruin and forest tree, is a useful plant, and often yields a winter meal to sheep, which are very fond of it. Many a homeless bird finds shelter amongst its luxuriant leaves,

> ' When autumn blasts are withering ;'

and its berries, ripening in April, thus afford food when heps and haws have passed away. Its leaves vary much in form : those on the lower branches are usually five-pointed, but higher up become more oval, and verge into one tapering point only. It has long been a question if the ivy injured the trees up which it climbed, and many botanists argue it does not. Its beauty pleads in its favour. The tenacity with which it adheres to the trees or old ruined wall it adorns, has led it to be considered an emblem of friendship. It is the badge of the clan Gordon.

But now let us hunt for flowers.

> ' The marsh is bleak and lonely. Scarce a flower
> Gleams in the waving grass ; .   .   .
> And the sole blossom which can glad the eye
> Is yon pale starwort nodding to the wind.'

This is the Michaelmas daisy, or sea starwort (*Aster Tripolium*), its pale lilac blossoms, with their yellow eye, growing in clusters, and rising on a stem some two feet high. Though attractive, it is not a pretty plant,—both leaves and flowers are of too pale and sickly a tint,—but we value it now, when brighter blossoms have passed from us.

> ' As cheerfulness lengthens the season of life,
>   So thou dost the season of flowers ;
> When the year waxeth old, thou with beauty art rife,
>   And thou smilest, though dark the sky lowers.'

And another starry blossom is smiling beneath the autumn sunshine, like a star of gold, from out the short grass, where the first falling leaves are withering. This is the hawkweed picris (*Picris hieracioides*), an abundant and common flower, with hairy leaves and stems, and yellow blossoms shaped like the dandelion, but lighter and more delicate. Very bright indeed are its starry petals gleaming from the hedge-side, where it sometimes grows as high as two or three feet ; and now that blossoms are becoming fewer each day, we welcome it with greater gladness. Caroline Bowles exclaims :

> ' Those few tall autumn flowers,
>   How beautiful they are !
> Than all that went before,
> Than all the summer store,
>   How lovelier far !'

Very pretty is the premorse scabious (*Scabiosa succisa*),

which now rears its round heads of bluish-purple blossoms. The flower grows on a tall, slender, hairy stem ;
its leaves are oblong, and also hairy, and yield a yellow
dye.

Here, too, is the ploughman's spikenard (*Inula conyza*),
with its dark green leaves, and dingy yellow flowers, still
in bloom. When bruised, it emits a strong odour of
camphor, from which it may have acquired its name ;
for camphor is supposed to be the spikenard of which
we read in the Song of Solomon. It is a plant growing
from two to three feet high, and is common in the south
of England, though somewhat rare in Scotland.

On the chalky hill we shall find a pretty little orchis,
called fragrant lady's tresses (*Neottia spiralis*), with
small, greenish-white blossoms, growing up a stem of a
few inches high, the leaves gathering in a tuft around
the root.

And near it we may gather the pretty field gentian
(*Gentiana campestris*), a purplish-lilac blossom ; and the
autumnal gentian (*Gentiana amarella*), which may sometimes be gathered in spring. It bears numerous blue
flowers, and may be distinguished from the field species
by the division of its corolla into five, instead of four
segments. There are five species of gentian, named
from Gentius, king of Illyria ; consequently Dryden has
called it

'The flower that bears inscribed the name of kings.'

As a wildling, it is decidedly uncommon ; but you must

often have been struck with its rich blue colouring adorning the garden border—

'The heaven-like blue of the little gentian'—

which, I am told, blossoms on the verge of frozen arctic regions—

'Living flowers that skirt eternal frost.'

Alas! our list of October blossoms is very short, and we have almost gathered all. Some of the flowers of previous months are bright and blooming still;

'And in the woods
A second blow of many flowers appears,
Flowers faintly tinged, and breathing no perfume.'

Peeping from amongst their overshadowing leaves, we may gather a second bloom of our loved violet; though it somewhat lacks the sweet fragrance of its earlier spring days, yet we greet it as a dear old friend.

'Under a hedge, all safe and warm,
Sheltered from boisterous wind and storm,
We violets lie,
With each small eye
Closely shut while the cold goes by.'

The wood sage (*Teucrium scorodonia*) may also still be gathered from the hedge-side. Though not a conspicuous flower, its spike of greenish-yellow blossoms and crumpled leaves render it rather a pretty plant. It has a bitter flavour, and is in some countries used in the place of hops to flavour beer.

Wood Sage—*Teucrium scorodonia.*

217

But October is more truly the month of fruits than flowers, as James Grahame tells us:

> ' Fruits, not blossoms, form the woodland wreath
> That circles autumn's brow : the ruddy haws
> Now clothe the half-leaved thorn ; the bramble bends
> Beneath its jetty load ; the hazel hangs
> With auburn branches.'

Yes! October is doubtless more the season of fruits. Though the spring blossoms have withered, their seeds have ripened 'neath the summer sun, and been dispersed over the earth, whilst our attention was wholly absorbed in gathering the fresh flowers as they appeared each month. Thus seedtime has been going on unheeded by us all the summer through, each blossom fulfilling the purpose for which it was designed by the Creator of all,

> ' Who, ere one flowery season fades and dies,
> Designs the blooming wonders of the next.'

Nature, grateful to the glad, genial sun and refreshing rain, is ever faithful.

> ' True to her trust, tree, herb, or reed
> She renders for each scattered seed,
> And to her Lord, with duteous heed,
> Gives large increase.
> Thus year by year she works unfee'd,
> And will not cease.'

While some seeds, like the feathery thistle-down, are wafted abroad by the faintest summer breeze, a stronger wind is required to scatter others, such as the ash, etc. Some are dependent on the humble birds for distribution ; others, again, wrapped in hard, thick covering, are allowed to rest where they fall, until the outer coat is decayed, and the matured seed finds its way into the soil, to spring up at its appointed time.   Thus we may ever

> 'Trace in nature's most minute design
> The signature and stamp of power divine.'

Oh! does not every flower, and seed, and fruit

> 'Speak unto our hearts of Him
> Who doeth all things well?'

And is it not wise to open the book of nature, and take

> 'The smallest flower
> That twinkles through the meadow grass, to serve
> For subject of a lesson?'

for in it we may ever read and learn,

> 'There lives and works
> A soul in all things, and that soul is God.'

XI.

NOVEMBER.

———◆———

'The melancholy days are come, the saddest of the year,
Of wailing winds and naked woods, and meadows brown and sere :
Heaped in the hollows of the grove, the withered leaves lie dead ;
They rustle to the eddying gust, and to the rabbit's tread.
The robin and the wren are flown, and from the shrub the jay ;
And from the wood-top calls the crow through all the gloomy day.

Where are the flowers, the fair young flowers, that lately sprung and stood
In brighter light and softer airs, a beauteous sisterhood ?
Alas ! they all are in their graves : the gentle race of flowers
Are lying in their lowly beds with the fair and good of ours.
The rain is falling where they lie ; but the cold November rain
Calls not from out the gloomy earth the lovely ones again.'

ERE and yellow leaves, with eddying sweep,' are falling, strewing the damp ground with their withering beauty, fast mouldering to decay. There is a quietness of repose all around, a sweet, tender sadness ; the vigour and eager growth of the year is over, and the earth is resting, as it were, from its labours ; the wind steals mournfully upon the ear, sighing through the leafless branches of the trees, where the homeless sparrow in vain seeks shelter; the melancholy rusty 'caw' of the rook is heard in the wood ;

> ' The little bird yet to salute the morn,
> Upon the naked branches sets her foot,
> The leaves now rising in the mossy root;
> And there a silly chirrupping doth keep,
> As though she fain would sing, yet fain would weep ;
> Praising fair summer, that so soon is gone,
> Or mourning winter too fast coming on.'

And thus November steals upon us, with its dews, and mists, and dimmer skies ; and there is little to tempt us out for a ramble where

> ' The dead leaves strew the forest walk,
> And withered are the pale wild-flowers.'

Yet still we may find some November blossoms to gather, and one or two we failed to notice in our later months, amongst which is the handsome common viper's bugloss (*Echium vulgare*), one of the most striking and stately of our wayside plants,—if less delicate than many, as elegant in shape as it is rich in hue. Its stems and leaves are spotted, and armed with strong, sharp bristles. These spots are likened to a snake's skin, and the seeds supposed to bear a resemblance to a viper's head ; and from these it has got its name— *Echium* being derived from the Greek signifying viper. It is a rough plant ; but very beautiful are its fine spikes of brilliant blue flowers which are closely set up the stem. It is peculiar in changing the colour of its blossoms from a deep red in the bud, to a purple in the opening flower, and a clear blue in the expanded or matured blossom. It flowers as early as June, but may be found as late as December growing by the way-side, the neglected field, or seaside beach.

'Here the blue bugloss paints the sterile soil,'
oft decking the cliff side or old wall with its beautiful blue blossoms, and rearing its head some two or three feet high. Here also we have the bugloss (*Anchusa sempervirens*), which you will observe has smaller clusters of flowers than the viper's bugloss, which are also a bright blue, nestling among the large, bristly leaves. A red dye is extracted from it, which was formerly used to paint the face. Its botanical name, *Anchusa*, from the Greek, signifies paint; hence the rhyme:

> ' Anchusa yields a ruddy stain,
> Which gives false colour to the face ;
> Be honest, maidens, and refrain
> From trifling with your natural grace.'

The greater knapweed (*Centaurea scabiosa*), which commences to bloom in August, is still in blossom : a handsome purple flower, very common in the corn-field or by the road-side :

> ' The iron-weed content to share
> The meanest spot the earth can spare.'

The knapweed is familiarly called iron-weed, from the hard cup on which its blossoms burst. Anne Pratt thinks that *knapweed* was probably *knob*-weed in former times. It greatly resembles the thistle, but may be recognised from it by possessing no prickles. The black knapweed (*Centaurea nigra*), which also flowers in July and August, is quite as common, but has smaller and duller blossoms, its leaves being rough and hairy. The juice of both plants is expressed for ink. To the same species belong the corn blue-bottle (*Centaurea cyanus*) and the star thistle (*Centaurea calcitrapa*), which we gathered in July.

The common ragwort (*Senecio Jacobæa*), though blooming first in July, may still be gathered; and handsome indeed are its bright yellow clusters of flowers, growing often two feet high. The leaves are toothed, the stem branched, and its golden blossoms emit a strong smell of honey. The marsh ragwort (*Senecio aquaticus*), which is usually found in wet pastures, greatly resembles it; the leaves, however, are not so divided, the purple flowers are larger, fewer, and scentless. The hoary ragwort (*Senecio tenuifolius*) is less common. It blossoms by the way-side, its erect stem being often three feet high. It has yellow flowers and pale green leaves, with a thick cottony down on the under surface. To the same species belongs the common groundsel (*Senecio vulgaris*), which we gathered in our February ramble, and which blossoms all the year through. Every child recognises it as a dainty morsel to pluck for the little caged canary at home.

Another yellow blossom that often lingers about the

hedge-side and waste place until December, is the
common nipplewort (*Lapsana communis*), a pale-flowered
long-stalked plant, with thin hairy leaves varying in
form. It has small yellow blossoms on branched straggling stems, and grows two, often three feet high. There
are, I believe, but two species. Alas! we have almost
come to an end of our November blossoms.

> ' The south wind searches for the flowers, whose fragrance late he
>   bore,
> And sighs to find them in the wood, and by the stream no more.'

Yet still a few autumn flowers are lingering.

> ' The rosy thrift,
> Though paler grown, since summer blessed the scene,

can still add beauty to our nosegay ; but

> ' The sea-lavender, whose lilac blooms
>   Drew from the saline soil a richer hue
>   Than when they grew on yonder towering cliff,
>   Quivers in flowerless greenness to the wind.'

We may chance to pick a stray crocus or two, doubtful wildlings, however ; and here is the ivy, still in its
flowering beauty and luxuriant foliage, wearing its bright,
evergreen tapestry o'er ' ruin grey,' or up to the highest
branches of the surrounding trees, clothing the naked
boughs with a freshness not their own, of which the
poet writes :

> ' Oh falsely they accuse me,
>   Who say I seek to check
> The growing saplings flourishing :—
>   I better love to deck

P

> The dead or dying branches
>   With all my living leaves ;
> 'Tis for the old and wither'd tree,
>   The ivy garland weaves.'

We may still find a pale Michaelmas daisy, a yellow hawkweed, a pretty little Mayweed, or spray of scabious ; and beautiful is

> ' The arbutus, heavy with its ruby fruit ;'

its waxen-tinted blossoms in drooping clusters, luxuriant stems, and bay-like leaves being more lovely than I can express.　Spenser tells us

> ' There is continual spring, and harvest there
>   Continual, both meeting at one time ;
> For both the boughs do laughing blossoms bear,
>   And with fresh colours deck the wanton prime,
>   And eke at once the heavy trees they climb,
>   Which seem to labour under their fruit's load.'

Ah ! we have still much beauty remaining to deck the November landscape, though the month to many seems so dull and dreary ; but to me each season comes fraught with its own peculiar attractions ; and

> ' I love to tread o'er autumn's sunny plains,
>   'Neath the brown foliage of its rich domains,
>   With mingled leaves of brownish green and gold,
>   Breathing their farewell tale to young and old.'

And I could almost exclaim with Keats :

> ' Let autumn bold,
> With universal tinge of sober gold,
>   Be all about me when I make my end.'

We may now find many very beautiful varieties of mosses, whether we wander in the wood or on the moor-side, some growing on the old wall, the rugged barks of the trees, or forming the soft carpet beneath our feet.

> ' Here a cluster of mosses too, tiny and rare,
> With leaves fine and glossy, and stems thin as hair,
> Not bearing gay flowers, but small cups instead,
> Which have each an extinguisher popped on its head.'

There are, I believe, about eight hundred different kinds of mosses; and though none are of any very great value to man, yet they are not wholly useless plants. A good dye is extracted from the club moss (*Lycopodium clavatum*). They protect seeds and even trees from winter cold, and afford shelter to myriads of the insect tribes. The reindeer moss (*Cladonia rangiferina*) is the winter food of the animal after which it is named; and in it we are given another evidence of the protective wisdom of the Great Creator, who in that cold and barren Lapland, destitute of fertile pastures, has provided a plentiful supply of this luxuriant moss, which neither cold nor snow can destroy. On the contrary, the snow shelters and protects the plant, which, though hidden from the sight, is quickly discovered by the natural instinct of the animal.

At this season especially, if the weather has been damp, we shall see many varieties of the fungi tribe, amongst which mushrooms and truffles are classed.

'Oh, look on the strange and the whimsical things,
   That among the wild fungi we find ;
And lichens and moss that like fairy-work springs ;
   If ye love not them all, ye are blind ;—
Ye are blind unto Nature's most glorious looks,
   If ye read not and love not her forest-born books.'

The species of fungi is so numerous that much time and space—which are not at our disposal—would be required to enter upon it ; but I dare say you have all observed in woods and shady places those leathery, spongy substances, some brightly coloured orange or red, a gay green, or pale lemon, growing on the decaying root of a tree, or any moist putrid matter, and have wondered what queer-looking unsightly things they were. Well, these are all fungi ; many smaller kinds, infesting wood and trees, are very destructive, and several are highly poisonous. Though you may deem them unsightly things, devoid of all beauty, yet if we examine the bright colouring of some, the singular variety and delicate formation of all, we cannot fail to be struck with admiration at the divine skill displayed even in the humblest vegetation of our beautiful world. Thus we see, my dear young friends, that an examination into the meanest work in the creation tends to make us both wiser and better. Coleridge tells us,

'He prayeth best, who loveth best
   All things both great and small ;
For the Great God, who loveth us,
   He made and loveth all.'

As I have said before, November seems to many a gloomy, melancholy month, wherein we ever

> ' Behold the emblem of our state,
>   In flowers that bloom and die.'

And an oppressive sadness steals over us in witnessing the hourly death of our beauteous summer flowers, for

> ' The yellow leaves around us fall,
>   Strewing the dewy glade :
>   A wondrous change is over all,
>   While loudly they to mortals call,
>   Like us you fade.'

Yes, it all appears sad and dismal enough ; but let us examine the leafless branch or bare stalk of the faded plant, and soon a reassuring calmness settles on our minds, a precious hope floods our souls ; for the still small voice of the undeveloped coming bud, that is *marked* on the frail stem that 'it is not dead but sleepeth,' whispers of the coming spring, the resurrection of flowers ; and though the words, 'Ye all do fade as a leaf,' are echoing through our hearts, faith points with steady hand 'to that inheritance incorruptible, and undefiled, that fadeth not away.'

XII.

DECEMBER.

———◆———

'There's not a blossom on the hill,
  There's not a leaf upon the tree :
The summer bird hath left its bough,
Bright child of sunshine, singing now
  In spicy lands beyond the sea.

The drooping year is in the wane,
  No longer floats the thistle down ;
The crimson heath is wan and sere,
The sedge hangs withering by the mere,
  And the broad fern is rent and brown.'

DECEMBER, the last month of the year—
'cold December, barrenness everywhere,'—has now arrived ; and with
regret, my dear young friends, I
come for our last gossip, our last
ramble together.

The blustering winds have swept the
withered leaves, and stripped the forest bare ;
the distant hills have already donned their
snowy caps ; the streams are bound with ice ; the dews
turned to hoar frosts ; the rook has returned to the nest

tree; the lesser birds are silent, and shivering in the leaf-less woods; and

> ' There's not a flower upon the lea ;
> The frost is on the pane.'

Yet we may still gather a bouquet of berries bright, from the transparent pearls of the mistletoe, to the dark purple fruit of the thorny burnet-rose, which will almost equal in beauty our summer nosegays of flowers.

Let us away this wintry morning, out into the clear, frosty air! A light shower of snow has fallen ; the feathery flakes, wheeling down from heaven, have covered the landscape with a thin carpet of dazzling whiteness, and have shrouded the forest branches with a sparkling fret-work of spotless purity; and oh! how exquisite is now the graceful mountain ash, with its rich clusters of bright coral berries peeping from amid the white drapery !

> ' In pearls and rubies rich the hawthorns show,
> While through the ice the crimson berries glow.'

The eglantine boasts that even in winter she has beauty :

> ' Though of both leaf and flower bereft,
> Such ornaments to me are left—
> Rich store of scarlet heps are mine.'

Here is the clear red fruit of the honeysuckle, to mix with the dark orange of the woody nightshade; the deep black of the bramble; the chocolate-brown of the ivy ; the crimson haws of the thorn ; the scarlet heps

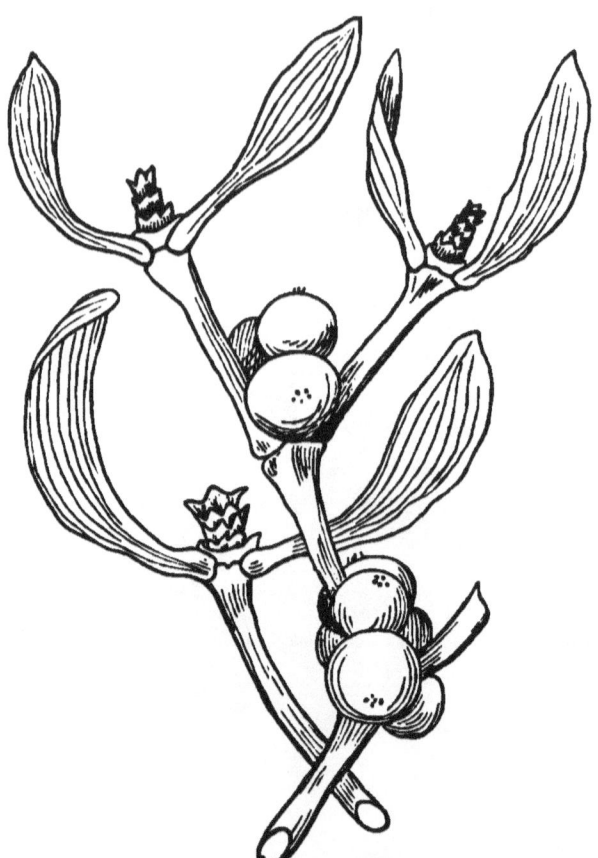

MISTLETOE—*Viscum album.*

of the dog-rose ; and the clustering, bright-gleaming berries of

> 'The holly, pointing to the moorland storm
>   Its hardy, fearless leaf.'

Beautiful indeed is the enduring foliage of the holly, its bright green leaves, in the cultivated kinds, mottled, or edged with pale yellow :

> 'When all the summer trees are seen
>   So bright and green,
> The holly's leaves a sober hue display,
>   Less bright than they ;
> But when the bare and wintry woods we see,
> What then so cheerful as the holly-tree ?'

Yes, bright and cheerful does the holly render our Christmas-time, brought in with the yew and ivy to decorate our churches, or, mixed with the mistletoe, to adorn our homes. It reminds one of the good old times when

> 'The damsel donned her kirtle sheen,
>   The hall was dressed with holly green ;
> Forth to the woods did merry men go,
> To gather in the mistletoe.
> Then open wide the baron's hall
> To vassal, tenant, serf, and all.'

You will all doubtless oft have admired the silvery, modest mistletoe (*Viscum album*), which cheers the woods and orchards of the south of England, and with its white, semi-transparent fruit, forms so beautiful a contrast to the scarlet-berried, spring-leaved holly. But few of you

may be aware of the interest attached to this plant,
which, though of no known use to man, excepting for
the birdlime made from the berries, is remarkable not
only as a parasite, but on account of the Druidical superstitions with which it is connected; and that a pagan
character seems still to be attached to it, is, I think,
proved by its not being permitted in church decorations,
though taken to ornament our houses. Now, the mistletoe, as I have said before, is a parasite, living on the
juices of other trees, to which its seeds have been by
chance or birds transferred and lodged in the bark.

The thistle tuft is indebted to the summer breeze for
the dispersion of its feathery seeds in the air, while the
stronger blast is required to scatter abroad the heavier
seeds of the ash, etc.; but it is to the humble and unin-
tentional ministry of a bird that the mistletoe owes its
dissemination; and thus we have a beautiful instance of
the innumerable means which our Father adopts in
spreading fertility, life, and beauty on our earth. The
berries of the mistletoe are the favourite food of many
birds, especially the missel-thrush, which has derived its
name from the fact. After a meal, they fly to the
nearest tree to rub and clean their bills from the clammy
substance which has issued from the fruit on which they
have been feasting, and thus the seed is conveyed, and
clings to the branch, stealing under the bark, and
speedily taking root. Any of you could easily try the
experiment for yourselves, by imitating this contrivance

of nature, and carefully crushing a ripe berry against the bark of an apple-tree early in spring. There are about twenty kinds of trees in England to which the mistletoe will attach itself, but on none more readily than the apple-tree, which it often destroys by sucking the sap, and thus robbing it of all nourishment : consequently the mistletoe is looked upon as an enemy in the orchard ; for

> ' It clingeth, it clingeth, and flourisheth still,
> And sucketh the juices, its own veins to fill,
> Of the stem which supports it : a parasite bold,
> That will never leave go, having once taken hold.'

But it was when found on the oak that the Druids held it in such sacred estimation, and associated it with such cruel practices and sacrifices as would make you shudder to hear of :

> ' The fearless British priests, under the aged oak
> Taking a milk-white bull, unstained with the yoke,
> And with an axe of gold, from that Jove-sacred tree
> The mistletoe cut down.'

It was gathered on the New-Year's morning with great pomp and solemnity. The chief Druid, clad in a white garment, ascended the tree, and cut down the mistletoe with a consecrated gold pruning hook, allowing it to fall into a pure white cloth, which was held suspended beneath by the other priests. It was considered a bad omen if a morsel chanced to touch the ground, and some terrible misfortune was anticipated. It was sold to the people, doubtless at high prices, as a sovereign remedy against all diseases, and a preservative against all dangers. Even as late as the seventeenth century, it was worn suspended around the neck as a safeguard against witches, and a cure for all sickness, so great were its imaginary virtues !

How thankful may *we* be that we live not in the days of such dark ignorance and superstition, that we only value the mistletoe for the lovely variety its pearly berries make in our Christmas decorations, and that its mystic influences extend no further than the sly kiss beneath its branches at

> ' The merry time of Christmas,
>   When young hearts slip the tether,
> And lips all merry, beneath the berry,
>   Close laughingly together ! '

Branches of the box-tree (*Buxus sempervirens*) are often mixed with the holly and mistletoe at this season, the leaf and general appearance of which must be familiar to you all. I believe there is but one species of box, which varies in size, as well as the tint of its leaves. The dwarf kind is remarkable for its fine glowing colour, which won it the title of 'sunny-coloured box.' The wood of the box is very valuable, being close-grained, hard, and heavy :

> ' Firm and smooth-grained, not easily
>   Warped or turned aside ;
> Like a stoic of old is the boxen tree,
>   Unmoved whate'er betide.'

It is used by the turner in making tops, chess-men, pegs, knife-handles, etc.; is valuable in the manufacture of mathematical instruments, flutes, combs, and, from its extreme hardness, is of especial service to the wood engraver. In olden times, sprigs of boxwood were carried at funerals, and Wordsworth tells us that a basinful was placed at the door of the house of mourning, and each person took a spray to throw into the grave:

> ' The basin of boxwood, just six months before,
> That stood on the table at Timothy's door;
> A coffin through Timothy's threshold had pass'd,
> One child did it bear, and that child was his last.'

The yew-tree (*Taxus baccata*) is also much used in our Christmas decorations, and very bright are its red, waxen-like berries, gleaming from amidst its dark, gloomy, evergreen foliage. It is often planted in church-yards as an emblem of immortality.

But when gathering our berries, let us see if no lingering flower is still to be found to gladden our sight.

> ' The wind-flower and the violet they perished long ago,
> And the wild rose and the orchis died amid the summer glow;
> But, on the hill, the golden rod, and the aster in the wood,
> And the yellow sunflower, by the brook, in autumn beauty stood,
> Till fell the frost from the clear cold heaven, as falls the plague on men,
> And the brightness of their smile was gone from upland glade and glen.'

Still, in spite of frost, we may chance to gather a stray corn marigold, with its gay golden dress, or the bright purple blossom of a knapweed, springing up from amongst the trailing branches of the bramble, which now exhibits many a reddish-green spray of leaves.

The bright yellow hawkweed picris, which we gathered in October, blooms yet as bright as ever, and will shed its beauty on the departing old year, as well as the dawn of the new.

See, here a timid shrinking daisy, that

> ' Golden tufte within a silver crowne,'

which some poets have named 'the robin of flowers,' raises its fair, frail head from out the snow, and gazes with sweet, innocent, glad face upturned to the frosty skies. Wordsworth wrote truly when he called it a

> ' Bright flower, whose home is everywhere !
> A pilgrim bold in nature's care ;
> And all the long year through, the heir
> Of joy or sorrow.'

And as true is Montgomery's assertion—

> ' The rose has but a summer's reign ;
> The daisy *never* dies.'

Turn your eyes upon yon desolate moorland, no longer glowing with the rich purple hue of our beautiful heaths ; no longer

> ' Thyme and marjoram are spreading
> Where thou art treading,
> And their sweet eyes for thee unclose ;'

but the exquisite glitter of the golden furze lingers there still, its half-expanded blossoms awaiting the first gleam of sunshine to spread their butterfly, wing-like flowers on the thorny stems.

> ' The prickly gorse, that shapeless and deformed,
>   And dangerous to the touch, has yet its bloom,
>   And decks itself with ornaments of gold.'

It is much prized in several of our continental countries, and is given a place in the richest greenhouses. It often attains a great height. The Rev. C. A. Johns, in his *Botanical Rambles*, speaks of having seen it so high, that when on horseback he could scarcely reach the top with his cane.

Alas! I fear we shall find few other blossoms to greet us at this season, for

> ' Flowers too bright, too sweet to last,
>   Drop all their leaves at winter's blast.'

And we must be content to think

> ' They'll soon return with genial spring,
>   More bright, more fair, more flourishing.'

Not only, however, in this our winter ramble do we miss the bright flowers of past seasons, but the full-tuned warblings of many birds no longer fill our souls with gladness. Ah !

> ' 'Tis sweet to see the opening flower
>   Spread its fair bosom to the sun ;
>   'Tis sweet to hear in vernal bower,
>   The thrush's earliest hymn begun ;'

yet no less sweet are the tender, glad trills of the bright-eyed little minstrel that warbles his song of trust

Q

and content from the leafless branch of a snow-clad
tree,—

> 'The bird whom man loves best,
>   The pious bird with the scarlet breast,
>   Our little English Robin:
>   The bird that comes about our doors,
>   When winter winds are sobbing.'

Who does not love the robin, which I have heard
called 'the daisy of birds,' as the daisy has been named
'the robin of flowers?' Who cannot remember how, as
children, we encouraged and tempted him with crumbs
to hop upon our window-sill,—a right welcome guest?

> 'Ever the first of all the birds
>   To hail the break of day;
> And sing what seems a little hymn,
>   To hail the sun's first ray.'

As we wander through the woods, let us pluck of the
many beautiful evergreens, mixing the box

> 'With the yew branch dark, and the mistletoe white,
> And holly and ivy leaves glossy and bright,
> With berries that gleam in the hearth's glad light,'

to deck our homes in honour of the coming Christmas,
in remembrance of the time when 'He came unto his
own, and his own received Him not;' when the heavenly
host proclaimed, 'Glory to God in the highest, and on
earth peace, good-will toward men.'

And now, my dear young friends, we have gathered
our December bouquet,—containing few flowers, it is
true, but rich in its glossy leaves and varied glittering

berries; and I must reluctantly bid you farewell. Our many sunny rambles are over; and to all I trust they have brought health, and strength, and better thoughts, to raise the mind from 'nature to nature's God.' I have striven—feebly enough it is true—to give you such information as the inquiring flower-lover would naturally crave, when plucking the way-side blossom, without taxing the memory with *hard names*, which have so often discouraged the young mind in its desire for knowledge regarding the beautiful wildlings of our lanes and meadows, moors and woodlands.

I have read that

> ' Flowers are the alphabet of angels, by which
> They write on hill and vale things unutterable ;'

and the idea is full of poetry and beauty. Shakespeare assures us that

> ' Fairies use flowers for their charactery ;'

and Horace Smith has called them

> ' Floral apostles, that in dewy splendour
> Weep without woe, and blush without a crime.'

Certain it is,

> ' Your voiceless lips, O flowers ! are living preachers,
> Each cup a pulpit, every leaf a book,
> Supplying to the fancy numerous teachers
> From loneliest nook.'

And thus the study of flowers brings us not only a fuller enjoyment in nature, a keener relish of the beauty

that surrounds us, but enlarged and deeper views of the
wisdom and goodness of our Almighty Father; and
therefore I would say—

> ' Blessed be God for flowers ;
> For the bright, gentle, holy thoughts that breathe
> From out their odorous beauty, like a wreath
> Of sunshine on life's hours.'

Let me trust, my dear young friends, that in these
our monthly gatherings

> ' The pure, sweet flowers of God'

have brought bright and happy thoughts to your minds,
added a new pleasure to your future ramble,—

> ' A cheerful faith, that all that we behold
> Is full of blessing ;'

that you may be able to exclaim, in the fulness of your
hearts—

> ' O Father, Lord !
> The All-beneficent !   I bless thy name
> That Thou hast mantled the green earth with flowers,
> Linking our hearts to nature !   By the love
> Of their wild blossoms our young footsteps first
> Into her deep recesses are beguiled.'

> .   .   .   .   .

> ' Receive
> Thanks, blessings, love, for these thy lavish boons,
> And most of all their heavenward influences,
> O Thou that gavest us flowers !'

GENERAL INDEX.

INDEX.

247

ILLUSTRATIONS.

MURRAY AND GIBB, EDINBURGH,

PRINTERS TO HER MAJESTY'S STATIONERY OFFICE.

LIST OF WORKS

PUBLISHED BY

JOHNSTONE, HUNTER, & CO., EDINBURGH.

MAGAZINES:—

The Christian Treasury; Containing Contributions from Ministers and Members of various Evangelical Denominations. Edited by HORATIUS BONAR, D.D. Super royal 8vo.

Monthly Parts,	·	·	·	·	£0 0	6
Weekly Numbers,	·	·	·	·	0 0	1

The Children's Hour; A Monthly Magazine for our Young Folks. Edited by M. H., Author of "Rosa Lindesay," etc. Crown 8vo. Beautifully Illustrated, · · · · 0 0 3

The Reformed Presbyterian Magazine; Containing Home and Missionary Intelligence relating to the Reformed Presbyterian Church in Scotland. Demy 8vo. Monthly, · · 0 0 4

J. H. & CO.'S SIXPENNY SERIES.

Super royal 32mo, cloth limp. Illustrated.

1. JEANIE HAY, THE CHEERFUL GIVER. And other Tales.
2. LILY RAMSAY; OR, HANDSOME IS WHO HANDSOME DOES. And other Tales.
3. ARCHIE DOUGLAS; OR, WHERE THERE'S A WILL THERE'S A WAY. And other Tales.
4. MINNIE AND LETTY; OR, THE EXPECTED ARRIVAL. And other Tales.
5. NED FAIRLIE AND HIS RICH UNCLE. And other Tales.
6. MR GRANVILLE'S JOURNEY. And other Tales.
7. JAMIE WILSON'S ADVENTURES. And other Tales.
8. THE TWO FRIENDS. And other Tales.
9. THE TURNIP LANTERN. And other Tales.
10. JOHN BUTLER; OR, THE BLIND MAN'S DOG. And other Tales.
11. CHRISTFRIED'S FIRST JOURNEY. And other Tales.
12. KATIE WATSON, THE CONTENTED LACEMAKER. And other Tales.
13. BIDDY, THE MAID OF ALL WORK.
14. MAGGIE MORRIS: A TALE OF THE DEVONSHIRE MOOR.

J. H. & CO.'S SHILLING PACKETS OF REWARD BOOKS.

Super Royal 32mo, in Illuminated Covers.

1. SHORT TALES TO EXPLAIN HOMELY PROVERBS. By M. H. A Series of Twelve Penny Books. Illustrated.
2. SHORT STORIES TO EXPLAIN BIBLE TEXTS. By M. H. A Series of Twelve Penny Books. Illustrated.

J. H. & CO.'S SHILLING PACKETS OF REWARD BOOKS—*Continued.*

    3. WISE SAYINGS, AND STORIES TO EXPLAIN THEM. By M. H. A Series of Twelve Penny Books. Illustrated.

    4. LITTLE TALES FOR LITTLE PEOPLE. A Series of Six Twopenny Books. Illustrated.

J. H. & CO.'S ONE SHILLING SERIES.

    Super royal 32mo, extra cloth, bevelled boards, Illustrated.

    1. THE STORY OF A RED VELVET BIBLE. By M. H.

    2. ALICE LOWTHER ; OR, GRANDMAMMA'S STORY ABOUT HER LITTLE RED BIBLE. By J. W. C.

    3. NOTHING TO DO ; OR, THE INFLUENCE OF A LIFE. By M. H.

    4. ALFRED AND THE LITTLE DOVE. By the Rev. F. A. Krummacher, D.D. And THE YOUNG SAVOYARD. By Ernest Hold.

    5. MARY M'NEILL ; OR, THE WORD REMEMBERED. A Tale of Humble Life. By J. W. C.

    6. HENRY MORGAN ; OR, THE SOWER AND THE SEED. By M. H.

    7. WITLESS WILLIE, THE IDIOT BOY. By the Author of "Mary Matheson," etc.

    8. MARY MANSFIELD ; OR, NO TIME TO BE A CHRISTIAN. By M. H.

    9. FRANK FIELDING ; OR, DEBTS AND DIFFICULTIES. By Agnes Veitch.

    10. TALES FOR "THE CHILDREN'S HOUR." By M. M. C.

    11. THE LITTLE CAPTAIN: a Tale of the Sea. By Mrs George Cupples.

    12. GOTTFRIED OF THE IRON HAND: a Tale of German Chivalry.

    13. ARTHUR FORTESCUE ; OR, THE SCHOOLBOY HERO. By Robert Hope Moncrieff.

    14. THE SANGREAL ; OR, THE HIDDEN TREASURE. By M. H.

    15. COCKERILL THE CONJUROR ; OR, THE BRAVE BOY OF HAMELN.

    16. JOTTINGS FROM THE DIARY OF THE SUN. By M. H.

    17. DOWN AMONG THE WATER WEEDS. By Mona B. Bickerstaffe.

J. H. & CO.'S EIGHTEENPENCE SERIES.

    Super royal 32mo, extra cloth, richly gilt sides and edges, Illustrated.

    1. SHORT TALES TO EXPLAIN HOMELY PROVERBS. By M. H.

    2. SHORT STORIES TO EXPLAIN BIBLE TEXTS. By M. H.

    3. ALFRED AND THE LITTLE DOVE. By the Rev. F. A. Krummacher, D.D. And WITLESS WILLIE, THE IDIOT BOY. By the Author of "Mary Matheson." etc.

    4. THE STORY OF A RED VELVET BIBLE: and HENRY MORGAN ; OR, THE SOWER AND THE SEED. By M. H., Editor of "The Children's Hour."

    5. ARTHUR FORTESCUE ; OR, THE SCHOOLBOY HERO. By Robert Hope Moncrieff. And FRANK FIELDING ; OR, DEBTS AND DIFFICULTIES. By Agnes Veitch.

    6. MARY M'NEILL ; OR, THE WORD REMEMBERED. By J. W. C. And other Tales.

    7. ALICE LOWTHER ; OR, GRANDMAMMA'S STORY ABOUT HER LITTLE RED BIBLE. By J. W. C. And other Tales.

    8. NOTHING TO DO ; OR, THE INFLUENCE OF A LIFE: and MARY MANSFIELD ; OR, NO TIME TO BE A CHRISTIAN. By M. H.

    9. BILL MARLIN'S TALES OF THE SEA. By Mrs George Cupples.

    10. GOTTFRIED OF THE IRON HAND. And other Tales.

    11. THE STORY OF THE KIRK: a Sketch of Scottish Church History. By Robert Naismith.

    12. THE HIDDEN TREASURE. And other Tales. By M. H.

    13. LITTLE TALES FOR LITTLE PEOPLE. By Various Authors.

    14. WISE SAYINGS, AND STORIES TO EXPLAIN THEM. By M. H.

J. H. & CO.'S HALF-CROWN SERIES.

Extra fcap. 8vo, handsomely bound in cloth.

1. ROSA LINDESAY, THE LIGHT OF KILMAIN. By M. H. Illustrated.
2. NEWLYN HOUSE, THE HOME OF THE DAVENPORTS. By A. E. W. Illustrated.
3. ALICE THORNE ; OR, A SISTER'S WORK. Illustrated.
4. LABOURERS IN THE VINEYARD. By M. H. Illustrated.
5. THE CHILDREN OF THE GREAT KING. By M. H. Illustrated.
6. LITTLE HARRY'S TROUBLES. By the Author of "Gottfried." Illustrated.
7. SUNDAY SCHOOL PHOTOGRAPHS. By the Rev. Alfred Taylor, Bristol, Pennsylvania.
8. WAYMARKS FOR THE GUIDING OF LITTLE FEET. By the Rev. J. A. Wallace.
9. THE DOMESTIC CIRCLE; OR, THE RELATIONS, RESPONSIBILITIES, AND DUTIES OF HOME LIFE. By the Rev. John Thomson. Illustrated.
10. SELECT CHRISTIAN BIOGRAPHIES. By the Rev. James Gardner, A.M., M.D. Illustrated.
11. OCEAN LAYS. Selected by the Rev. J. Longmuir, LL.D. Illustrated.
12. WILBERFORCE'S PRACTICAL VIEW OF CHRISTIANITY. Complete Edition.
13. COMMUNION SERVICES, ACCORDING TO THE PRESBYTERIAN FORM. By the Rev. J. A. Wallace.
14. ATTITUDES AND ASPECTS OF THE DIVINE REDEEMER. By Rev. J. A. Wallace.
15. THE REDEEMER AND THE REDEMPTION. By the Rev. Alex. S. Patterson, D.D.
16. A PASTOR'S LEGACY. Edited by the Rev. J. A. Wallace.
17. JAMES NISBET ; A STUDY FOR YOUNG MEN. By the Rev. J. A. Wallace.
18. NOBLE RIVERS, AND STORIES CONCERNING THEM. By Anna J. Buckland. Illustrated.

J. H. & CO.'S THREE SHILLING SERIES.

Extra fcap. 8vo, richly gilt sides and edges.

1. MISS MATTY ; OR, OUR YOUNGEST PASSENGER. And other Tales. Illustrated.
2. HORACE HAZELWOOD ; OR, LITTLE THINGS. And other Tales. Illustrated.
3. ROSA LINDESAY, THE LIGHT OF KILMAIN. By M. H. Illustrated.
4. NEWLYN HOUSE, THE HOME OF THE DAVENPORTS. By A. E. W. Illustrated.
5. ALICE THORNE ; OR, A SISTER'S WORK. Illustrated.
6. LABOURERS IN THE VINEYARD. By M. H. Illustrated.
7. LITTLE HARRY'S TROUBLES. By the Author of "Gottfried." Illustrated.
8. THE CHILDREN OF THE GREAT KING. By M. H. Illustrated.
9. THE DOMESTIC CIRCLE ; OR, THE RELATIONS, RESPONSIBILITIES, AND DUTIES OF HOME LIFE. By the Rev. John Thomson., Illustrated.
10. SUNDAY SCHOOL PHOTOGRAPHS. By the Rev. Alfred Taylor, Bristol.
11. WAYMARKS FOR THE GUIDING OF LITTLE FEET. By the Rev. J. A. Wallace.
12. SELECT CHRISTIAN BIOGRAPHIES. By the Rev. James Gardner, A.M., M.D Illustrated.
13. CARDIPHONIA ; OR, THE UTTERANCE OF THE HEART. In a Series of Letters. By John Newton. A New Edition, bevelled boards, cut edges.
14. FOUND AFLOAT. By Mrs George Cupples. And other Tales. Illustrated.
15. JAMES NISBET ; A STUDY FOR YOUNG MEN. By the Rev. J. A. Wallace.
16. THE WHITE ROE OF GLENMERE. By Mona B. Bickerstaffe. And other Tales. Illustrated.
17. NOBLE RIVERS, AND STORIES CONCERNING THEM. By Anna J. Buckland. Illustrated.

J. H. & CO.'S FIVE SHILLING SERIES.
Bound in cloth, bevelled boards, richly gilt sides and edges.

1. THE CHILDREN'S HOUR ANNUAL. First Series. 656 pp. Extra fcap. 8vo. Illustrated.
2. THE CHILDREN'S HOUR ANNUAL. Second Series. 640 pp. Extra fcap. 8vo. Illustrated.
3. SKETCHES OF SCRIPTURE CHARACTERS. By the Rev. Andrew Thomson, D.D. Crown 8vo. Illustrated.
4. STARS OF EARTH ; OR, WILD FLOWERS OF THE MONTHS. By Leigh Page. Crown 8vo. With Original Illustrations by the Author.
5. ELIJAH ; THE DESERT PROPHET: A Biography. By the Rev. H. T. Howat. Crown 8vo. Illustrated.

Afflicted's Refuge (The) ; or, Prayers adapted to various Circumstances of Distress. Fcap. 8vo, cloth,   £0 2 6

Alfred and the Little Dove. By F. A. Krummacher, D.D. And the Young Savoyard. By Ernest Hold. Translated from the German by a Lady. Royal 32mo, cloth, Illustrated, -   0 1 0

Alice Lowther ; or, Grandmamma's Story about her Little Red Bible. By J. W. C., Author of "Mary M'Neill," etc. Royal 32mo, cloth, Illustrated, -   0 1 0

Alice Thorne ; or, A Sister's Work.
Extra fcap. 8vo, cloth, Illustrated, -   0 2 6
———————— Extra cloth, gilt edges, -   0 3 0

Archie Douglas ; or, Where there's a Will there's a Way. And other Tales. Super Royal 32mo, cloth, Illustrated, -   0 0 6

Arthur Fortescue ; or, The Schoolboy Hero.
By Robert Hope Moncrieff. Royal 32mo, cloth, Illustrated, -   0 1 0

Authorised Standards of the Free Church of Scotland:
Being the Westminster Confession of Faith and other Documents. *Published by Authority of the General Assembly.* Demy 12mo, cloth limp, -   0 1 3
———————— Cloth boards, -   0 1 6
——— Superior Edition, Printed on Superfine Paper, extra cloth, bevelled boards, antique, -   0 2 6
———————— Full calf, lettered, antique, -   0 5 0

Biddy, the Maid of All Work.
Super Royal 32mo, cloth, Illustrated, -   0 0 6

Bill Marlin's Tales of the Sea.
By Mrs George Cupples. Super royal 32mo, extra cloth, gilt edges, Illustrated, -   0 1 6

Brodie (Rev. James, A.M.) The Antiquity and Nature of Man:
A Reply to Sir Charles Lyell's Recent Work. Extra fcap. 8vo, cloth,   0 2 6
——— Papers Offered for Discussion at the Meeting of the British Association at Dundee. Extra fcap. 8vo, boards, -   0 1 0
——— The Rational Creation: An Inquiry into the Nature and Classification of Rational Creatures, and the Government which God exercises over them. Crown 8vo, cloth, -   0 5 0
——— An Inquiry into the Apocalypse, with an Endeavour to ascertain our present Position on the Chart of Time. Royal 8vo, sewed,   0 2 0

Brodie (Rev. James, A. M.) Memoir of Annie M'Donald Christie,
a Self-taught Cottager. Demy 18mo, cloth, -   -   - £0   1   6

Buckland (Anna J.) Noble Rivers, and Stories concerning Them.
Extra fcap. 8vo, cloth, with Illustrations, -   -   -   0   2   6
———— Extra cloth, gilt edges,   -   -   -   -   0   3   0

Burns (Rev. George, D.D.) Prayers for the Use of Sabbath Schools.
18mo, sewed,   -   -   -   -   -   -   -   0   0   4

Catechisms—

THE ASSEMBLY'S SHORTER CATECHISM; with References to the
Scripture Proofs. Demy 18mo, stitched, -   -   0   0   0½

THE ASSEMBLY'S SHORTER CATECHISM; with (*Italicised*) Proofs from
Scripture at full length; also with Additional Scripture References,
selected from Boston, Fisher, M. Henry, Paterson, Vincent, and
others. Demy 18mo, stitched, -   -   0   0   1

THE ASSEMBLY'S LARGER CATECHISM; with (*Italicised*) Proofs from
Scripture at full length. Demy 12mo, sewed,   -   0   0   6

CATECHISM OF THE EVIDENCES OF REVEALED RELIGION, with a few
Preliminary Questions on Natural Religion. By William Ferrie,
D.D., Kilconquhar. 18mo, stitched,   -   0   0   4

CATECHISM ON BAPTISM: in which are considered its Nature, its
Subjects, and the Obligations resulting from it. By the late
Henry Grey, D.D., Edinburgh. 18mo, stitched, -   0   0   6

THE CHILD'S FIRST CATECHISM. 48mo, stitched,   -   0   0   0½

SHORT CATECHISM FOR YOUNG CHILDREN. By the Rev. John Brown,
Haddington. 32mo, stitched,   -   0   0   0½

PLAIN CATECHISM FOR CHILDREN. By the Rev. Matthew Henry.
18mo, stitched,   -   -   -   0   0   1

FIFTY QUESTIONS CONCERNING THE LEADING DOCTRINES AND DUTIES
OF THE GOSPEL; with Scripture Answers and Parallel Texts. For
the use of Sabbath Schools. 18mo, stitched,   -   0   0   1

FORM OF EXAMINATION BEFORE THE COMMUNION. Approved by the
General Assembly of the Kirk of Scotland (1592), and appointed
to be read in Families and Schools; with Proofs from Scripture
(commonly known as "Craig's Catechism"). With a Recommendatory Note by the Rev. Dr Candlish, Rev. Alexander Moody
Stuart, and Rev. Dr Horatius Bonar. 18mo, stitched,   -   0   0   1

THE MOTHER'S CATECHISM; being a Preparatory Help for the Young
and Ignorant, to their easier understanding The Assembly's
Shorter Catechism. By the Rev. John Willison, Dundee. 32mo,
stitched,   -   -   -   -   0   0   1

WATTS' (DR ISAAC) JUVENILE HISTORICAL CATECHISMS OF THE OLD
AND NEW TESTAMENTS; with Numerous Scripture References, and
a Selection of Hymns. Demy 18mo, stitched,   -   0   0   1

A SCRIPTURE CATECHISM, Historical and Doctrinal, for the use of
Schools and Families. By John Whitecross, Author of "Anec-
dotes on the Shorter Catechism," etc. 18mo, stitched,   0   0   1

A SUMMARY OF CHRISTIAN DOCTRINE AND DUTIES; being the Westminster Assembly's Shorter Catechism, without the Questions,
with Marginal References. Fcap. 8vo, stitched, -   0   0   1

Children of the Great King (The): A Story of the Crimean War.
By M. H., Editor of "The Children's Hour." Extra fcap. 8vo,
cloth, with Illustrations,   -   -   -   0   2   6
———— Cloth extra, gilt edges,   -   -   -   0   3   0

Children's Hour (The) Annual. First Series.
656 pp., and upwards of 50 Illustrations. Extra fcap. 8vo, cloth, gilt edges,   -   -   -   -   -   -   £0   5   0

—— Second Series.
640 pp., and upwards of 70 Illustrations. Extra fcap. 8vo, cloth, gilt edges,   -   -   -   -   -   -   0   5   0

Children's Hour (The) Series of Gift Books.
1. MISS MATTY; OR, OUR YOUNGEST PASSENGER. And other Tales.
2. HORACE HAZELWOOD. And other Tales.
3. FOUND AFLOAT. And other Tales.
4. THE WHITE ROE OF GLENMERE. And other Tales.
Extra fcap. 8vo, cloth, gilt sides and edges, Illustrated—each   0   3   0

Christfried's First Journey. And other Tales.
Super royal 32mo, cloth, Illustrated,   -   -   -   0   0   6

Christian Treasury (The) Volumes 1845 to 1860.
16 Volumes, royal 8vo, cloth—each   -   -   -   0   5   0
A complete Set will be forwarded to any part of the country, carriage paid, on receipt of £3, 3s.

—— Volumes 1861, 1862, 1863, 1864, 1865, and 1866.
Super royal 8vo, cloth, green and gold—each   -   -   0   6   6

Cockerill the Conjurer; or, The Brave Boy of Hameln. By the
Author of "Little Harry's Troubles." Super royal 32mo, cloth, bevelled boards, Illustrated,   -   -   -   -   0   1   0

Confession of Faith (The) agreed upon at the Assembly of
Divines at Westminster. Complete Edition, with the *Italics* of the elegant Quarto Edition of 1658 restored. (Authorised Edition.)
Demy 12mo, cloth limp,   -   -   -   -   0   1   0
—————— Cloth boards,   -   -   -   -   -   0   1   3
—— Superior Edition, Printed on Superfine Paper, extra cloth, bevelled boards, antique,   -   -   -   -   -   0   2   6
—————— Full calf, lettered, antique,   -   -   -   0   5   0

Dill (Edward Marcus, A.M., M.D.) The Mystery Solved: or,
Ireland's Miseries: Their Grand Cause and Cure. Fcap. 8vo, cloth,   0   2   6

—— The Gathering Storm; or, Britain's Romeward Career: A
Warning and Appeal to British Protestants. Fcap. 8vo, cloth,   0   1   0

Down among the Water Weeds; or Marvels of Pond Life.
By Mona B. Bickerstaffe. Super royal 32mo, cloth, bevelled boards, with numerous Illustrations,   -   -   -   -   0   1   0

Family Prayers for Working Men.
By Ministers of Various Evangelical Denominations. Edited with a Preface by the Rev. John Hall, D.D., Dublin. Extra fcap. 8vo—
Common Edition, stiffened boards,   -   -   -   0   0   6
—————— limp cloth,   -   -   -   -   0   0   9
Fine Edition, bevelled boards,   -   -   -   0   1   6

Forbes (Rev. Robert, A.M.) Procedure in the Inferior Courts of
the Free Church of Scotland. With Appendix, embracing a Ministerial Manual; with Forms and Documents.
(Third Edition in Preparation.)

Found Afloat. By Mrs George Cupples. And other Tales.
Extra fcap. 8vo, with Illustrations. Cloth, gilt edges,   -   0   3   0

Frank Fielding; or, Debts and Difficulties.
A Story for Boys. By Agnes Veitch, Author of "Woodruffe," etc. Royal 32mo, bevelled boards, - - - - £0 1 0

Gardner (Rev. James, A.M., M.D.) Select Christian Biographies.
Extra fcap. 8vo, cloth, with Illustrations, - - - 0 2 6
———— Cloth extra, gilt edges, - - - - 0 3 0

Gottfried of the Iron Hand: A Tale of German Chivalry.
By the Author of "Little Harry's Troubles." Royal 32mo, cloth, Illustrated, - - - - - - 0 1 0

Habit; with Special Reference to the Formation of a Virtuous
Character. An Address to Students. By Burns Thomson. With a Recommendatory Note by the late Professor Miller. 18mo, Second Edition, revised, - - - - - 0 0 2

Henry Morgan; or, The Sower and the Seed.
By M. H., Editor of "The Children's Hour." Royal 32mo, cloth, Illustrated, - - - - - - 0 1 0

Hidden Treasure (The). And other Tales.
By M. H., Editor of "The Children's Hour." Super royal 32mo, cloth extra, gilt edges, Illustrated, - - - 0 1 6

Horace Hazelwood; or, Little Things.
By Robert Hope Moncrieff. And other Tales. Extra fcap. 8vo, with Illustrations, cloth, gilt edges, - - - 0 3 0

Howat (Rev. H. T.) Elijah; the Desert Prophet. A Biography.
Crown 8vo, cloth, gilt edges, with Illustrations. - - 0 5 0

———— Sabbath Hours: A Series of Meditations on Gospel Themes.
Extra fcap. 8vo, cloth, - - - - - 0 3 6

Hunter (James J.) Historical Notices of Lady Yester's Church
and Parish, Edinburgh. Compiled from Authentic Sources. Extra fcap. 8vo, cloth, Printed on Toned Paper, - - 0 2 0

Hymns for the Use of Sabbath Schools and Bible Classes.
Selected by a Committee of Clergymen. Royal 32mo, sewed, - 0 0 3

Jamie Wilson's Adventures. And other Tales.
Super royal 32mo, cloth, Illustrated, - - - 0 0 6

Jeanie Hay, the Cheerful Giver. And other Tales.
Super royal 32mo, cloth, Illustrated, - - - 0 0 6

John Butler; or, The Blind Man's Dog. And other Tales.
Super royal 32mo, cloth, Illustrated, - - - 0 0 6

Jottings from the Diary of the Sun.
By M. H., Editor of "The Children's Hour." Super royal 32mo, cloth, bevelled boards, Illustrated, - - - 0 1 0

Katie Watson. And other Tales.
Super royal 32mo, cloth, Illustrated, - - - 0 0 6

Labourers in the Vineyard; or, Dioramic Scenes from the Lives
of Eminent Christians. With Recommendatory Preface, by the Rev. A. K. H. Boyd, D.D., St Andrews. Extra fcap. 8vo, cloth, with Illustrations, - - - - - 0 2 6
———— Cloth extra, gilt edges, - - - - 0 3 0

Lily Ramsay; or, Handsome Is who Handsome Does. And other
Tales. Super royal 32mo, cloth, Illustrated, -   -   -   £0 0 6

Little Captain (The): A Tale of the Sea.
By Mrs George Cupples. Royal 32mo, cloth, Illustrated,   -   0 1 0

Little Harry's Troubles: A Story of Gipsy Life.
By the Author of " Gottfried of the Iron Hand." Extra fcap. 8vo,
cloth, with Illustrations, -   -   -   -   -   0 2 6
————— Cloth extra, gilt edges,   -   -   -   -   0 3 0

Little Tales for Little People. A Series of Twopenny Reward
Books. By Various Authors. With Illustrations.

    1. John Butler; or, The Blind Man's Dog.
       Rosedale Villa ; or, The Grey Parrot.
    2. Asta Von Flotow ; or, "The Cuckoo's Warning."
       The Two Blackbirds ; or, Jealousy.
    3. Charles Harley ; or, The Bag of Marbles.
       Stanley Hollins ; or, The Spider and the Silkworms.
    4. Christfried's First Journey.
    5. Katie Watson, the Contented Lacemaker.
       The Twin Brothers : A Ragged School Reminiscence.
    6. Willie Smith : A Scottish Story.
       Alice and Fanny ; or, Disobedience Punished.

       Done up in a neat Illuminated Packet. Super royal 32mo,   0 1 0
       In cloth extra, gilt edges,   -   -   -   -   0 1 6

M'Donald (Rev. John, Calcutta). The Suffering Saviour.
With a Preface, by the late Rev. W. K. Tweedie, D.D. Royal 32mo,
limp cloth,   -   -   -   -   -   -   -   0 0 6

Maggie Morris: A Tale of the Devonshire Moor.
Super royal 32mo, cloth, Illustrated,   -   -   -   0 0 6

Mary M'Neill; or, The Word Remembered.
A Tale of Humble Life. By J. W. C., Author of " Alice Lowther,"
etc. Royal 32mo, cloth, Illustrated,   -   -   -   0 1 0

Mary Mansfield; or, No Time to be a Christian.
By M. H., Editor of " The Children's Hour." Royal 32mo, cloth,
Illustrated,   -   -   -   -   -   -   -   0 1 0

Meikle (Rev. James, D.D.) Coming Events.
An Inquiry regarding the Three Prophetical Numbers of the last
Chapter of Daniel. Extra fcap. 8vo, cloth,   -   -   0 2 6

—— **The Battle of Armageddon; and its Results.**
An Exposition of the Sixth and Seventh Vials of the Apocalypse.
And also an Inquiry regarding the Commencement of the 1260
Symbolical Days. Crown 8vo, cloth,   -   -   -   0 3 6

—— **The Edenic Dispensation.**
Fcap. 8vo, cloth,   -   -   -   -   -   0 3 6

—— **The Nature of the Mediatorial Dispensation.**
Crown 8vo, cloth,   -   -   -   -   -   0 3 6

—— **The Administration of the Mediatorial Dispensation.**
Crown 8vo, cloth,   -   -   -   -   -   0 3 6

Miller (Rev. James N.) Prelacy Tried by the Word of God.
With an Appendix on the Prelatic Argument from Church History.
Fcap. 8vo, limp cloth, - - - - - - £0 1 0

Miller (Professor James). Physiology in Harmony with the Bible,
respecting the Value and Right Observance of the Sabbath. Royal
32mo, limp cloth, - - - - - - 0 0 6

Minnie and Letty; or, the Expected Arrival. And other Tales.
Super royal 32mo, cloth, Illustrated, - - - 0 0 6

Miss Matty; or, Our Youngest Passenger.
By Mrs George Cupples. And other Tales. Extra fcap. 8vo, with
Illustrations, cloth, gilt edges, - - - - 0 3 0

Mr Granville's Journey. And other Tales.
Super royal 32mo, cloth, Illustrated, - - - 0 0 6

MUSIC—

The Treasury Hymnal, a Selection of Part Music, in the
ordinary Notation, with Instrumental Accompaniment; the
Hymns selected from Dr Bonar's "Hymns of Faith and Hope."
The Letter Note Method of Musical Notation, by permission of
Messrs Colville and Bentley, is added as a help to young singers.

No. 1.	FORWARD,	Old Melody.
	A BETHLEHEM HYMN,	Arranged from Mozart.
2.	THE FRIEND,	Haydn.
	LOST BUT FOUND,	Pleyel.
3.	A LITTLE WHILE,	Adapted from Mendelssohn.
	A STRANGER HERE,	Pleyel.
4.	THE BLANK,	Pleyel.
	THE NIGHT AND THE MORNING,	Adapted from Rode.
5.	THE CLOUDLESS,	Haydn.
	THE SUBSTITUTE,	Haydn.
6.	THY WAY, NOT MINE,	Altered from Pleyel.
	REST YONDER,	Steibelt.
7.	EVER NEAR,	German Melody.
	QUIS SEPARABIT,	Beethoven.
8.	ALL WELL,	Haydn.
	DISAPPOINTMENT,	Haydn.
	CHILD'S PRAYER,	Weber.
9.	GOD'S ISRAEL,	Atterbury.
	THE ELDER BROTHER,	Beethoven.
	DAY SPRING,	German Melody.
10.	THE NIGHT COMETH,	Venetian Melody.
	HOW LONG,	Mendelssohn.
11.	THE TWO ERAS,	Spohr.
	THE SHEPHERD'S PLAIN,	Whitaker.
12.	BRIGHT FEET OF MAY,	Whitaker.
	HEAVEN AT LAST,	Clementi.

The above in stiffened wrapper. Super royal 8vo, - 0 1 0
Single Nos.—each - - - - 0 0 1

A Graduated Course of Elementary Instruction in Singing
ON THE LETTER NOTE METHOD (*by means of which any difficulty
of learning to Sing from the common Notation can be easily over-
come*) in 26 Lessons. By David Colville and George Bentley.
Royal 8vo, in wrapper, - - - - 0 1 0
———— Cloth, - - - 0 1 6

The Pupil's Hand-Book, being the Exercises contained in
the foregoing Work. For the use of Classes and Schools. 2 Parts,
sold separately—each - - - - - 0 0 3

MUSIC—*Continued.*

A Junior Course of Instruction in Singing on the Letter Note Method. Arranged progressively on the same plan as the Graduated Course, and specially for the use of Schools and Junior Classes. Nos. 1 and 2 published—each · · · £0 0 1

An Elementary Course of Practice in Vocal Music, with numerous Tables, Diagrams, etc., and copious Examples of all the usual Times, Keys, and Changes of Keys. For use in connection with any method of Solmization. By David Colville. Complete in 2 Parts—each · · · · 0 0 4

Colville's Choral School; A Collection of Easy Part Songs, Anthems, etc., in Vocal Score, for the use of Schools and Singing Classes. Arranged progressively, and forming a complete Course of Practice in Vocal Music. In 20 Parts—each · · 0 0 4

Choral Harmony, in Vocal Score, for the use of Choral Societies, Classes, Schools, etc. Edited by David Colville. 100 Nos. published. List on page 13. Royal 8vo—each · · 0 0 1

—— Nos. 1 to 16, in Illuminated wrapper, stiffened, · 0 1 0
——— Nos. 17 to 34, in Illuminated wrapper, stiffened, · 0 1 0
—— Nos. 35 to 50, in Illuminated wrapper, stiffened, · 0 1 0
—— Vol. I. (50 Nos.), cloth boards, · · · 0 4 0
—— Vol. II. (50 Nos.), cloth boards, · · · 0 4 0

Naismith (Robert). The Story of the Kirk.
A Sketch of Scottish Church History for Young Persons. Super royal 32mo, cloth extra, gilt edges, · · · · 0 1 6

Ned Fairlie and his Rich Uncle. And other Tales.
Super royal 32mo, cloth, Illustrated, · · · 0 0 6

Newlyn House, the Home of the Davenports.
By E. A. W. Extra fcap. 8vo, cloth, with Illustrations. · 0 2 6
——— Extra cloth, gilt edges, · · · · 0 3 0

Newton (John). Cardiphonia; or, The Utterance of the Heart.
Extra fcap. 8vo, cloth, · · · · · 0 3 0

Nothing to Do; or, The Influence of a Life.
By M. H., Editor of "The Children's Hour." Royal 32mo, cloth, Illustrated, · · · · · · 0 1 0

Ocean Lays; or, the Sea, the Ship, and the Sailor.
A Selection of Poetry. By the Rev. J. Longmuir, LL.D. Royal 16mo, cloth, Illustrated, · · · · 0 2 6

Patterson (Alexander S., D.D.) The Redeemer and the Redemption.
A Series of Sacramental Discourses. Extra fcap. 8vo, cloth, · 0 2 6

Red Velvet Bible (The Story of a).
By M. H., Editor of "The Children's Hour." Royal 32mo, cloth, Illustrated, · · · · · · 0 1 0

Rosa Lindesay, the Light of Kilmain.
By M. H., Editor of "The Children's Hour." Extra fcap. 8vo, with Illustrations, · · · · 0 2 6
——— Cloth extra, gilt edges, · · · · 0 3 0

Sangreal (The); or, The Hidden Treasure.
By M. H., Editor of "The Children's Hour." Super royal 32mo, cloth, bevelled boards, Illustrated, -   -   -   -   £0 1 0

Short Stories to Explain Bible Texts.
By M. H., Editor of "The Children's Hour." With Illustrations.

1. Minnie and Letty.
2. Willie Lewis and his Schoolfellows.
3. Discontented Susy.
4. Charlie Grant and his Sister Nina.
5. Ned Fairlie and his Rich Uncle.
6. Annie Ross; the Little Housekeeper.
7. Little Jephy; the Adopted Child.
8. Harry Westbrook's Visit to Grandpapa.
9. Mr Granville's Journey.
10. Stella Howard and her Morning Calls.
11. Bertie and Ethel; or, Self-Denial.
12. Little Milly and her Half-Crown.

Done up in a neat Illuminated Packet, 32mo, -   -   £0 1 0
In cloth extra, gilt edges,   - - - - - -   0 1 6

Short Tales to Explain Homely Proverbs; a Series of Reward Books.
By M. H., Editor of "The Children's Hour." With Illustrations by Charles A. Doyle.

1. Who Gives Quickly Gives Twice.
2. Short Accounts make Long Friends.
3. Evil Communications Corrupt Good Manners.
4. Forgive and Forget.
5. Handsome Is who Handsome Does.
6. Better Late than Never.
7. Do as You Would be Done by.
8. A Stitch in Time saves Nine.
9. Where there's a Will there's a Way.
10. All is not Gold that Glitters.
11. Waste not, Want not.
12. There is no Place like Home.

Done up in a neat Illuminated Packet, 32mo, -   -   £0 1 0
In cloth extra, gilt edges,   - - - - -   0 1 6

Stars of Earth; or, Wild Flowers of the Months.
By Leigh Page. With upwards of 50 original Illustrations by the Author. Crown 8vo, cloth extra, gilt edges, -   -   -   0 5 0

Steele (James). A Manual of the Evidences of Christianity.
Chiefly intended for Young Persons. 18mo, cloth, -   -   0 1 0

Stocking Knitter's Manual (The).
A Companion for the Work Table. By Mrs George Cupples. Extra fcap. 8vo, Illuminated Cover, -   -   -   0 0 6

Story of Daniel (The).
From the Original of the late Professor Gaussen. By the Author of "The World's Birthday." Extra fcap. 8vo, cloth, with Illustrations,
(In Preparation.)

Sunday School Photographs.
By the Rev. Alfred Taylor, Bristol, Pennsylvania. With Introduction by John S. Hart, LL.D., Philadelphia, U.S. Extra fcap. 8vo, cloth, printed on toned paper, -   -   -   -   0 2 6
———— Extra cloth, gilt edges, -   -   -   0 3 0

Tales for the Children's Hour. By M. M. C.
Royal 32mo, cloth, Illustrated, -   -   -   -   0 1 0

Thomson (Rev. Andrew, D.D.) Sketches of Scripture Characters.
Crown 8vo, cloth extra, gilt edges, with Illustrations, -   -   0 5 0

Thomson (Rev. John.) The Domestic Circle; or, The Relations,
Responsibilities, and Duties of Home Life. Extra fcap. 8vo, cloth,   0 2 6
———— Extra cloth, gilt edges, -   -   -   0 3 0

Thoughts on Intercessory Prayer. By a Lady.
Royal 32mo, limp cloth,   -   -   -   -   -   £0   0   6

Turnip Lantern (The). And other Tales.
Super royal 32mo, cloth, Illustrated,   -   -   0   0   6

Two Friends (The). And other Tales.
Super royal 32mo, cloth, Illustrated,   -   -   0   0   6

Tytler (C. E. Fraser). The Seals and Roll of St John.
Demy 8vo, cloth,   -   -   -   -   -   0   7   6

———— The Structure of Prophecy and of the Apocalypse.
Demy 8vo, cloth,   -   -   -   -   -   0   3   6

———— The Apocalypse.
Demy 8vo, cloth, -   -   -   -   -   0   5   0

Wallace (Rev. J. A.) Attitudes and Aspects of the Divine Redeemer.
Extra fcap. 8vo, cloth, printed on toned paper,   -   -   0   2   6

———— Pastoral Recollections. Third Series. 1853—63.
Extra fcap. 8vo, cloth, printed on toned paper,   -   -   0   2   6

———— A Pastor's Legacy. Being Brief Extracts from the MSS.
of the late Rev. A. B. Nichol, Galashiels. With Introductory
Notice. Extra fcap. 8vo, cloth, printed on toned paper,   -   0   2   6

———— Communion Services According to the Presbyterian Form.
Extra fcap. 8vo, cloth, printed on toned paper,   -   -   0   2   6

———— Waymarks for the Guiding of Little Feet.
Extra fcap. 8vo, cloth, printed on toned paper,   -   -   0   2   6
———— Extra cloth, gilt edges,   -   -   -   0   3   0

Wallace (Rev. J. A.) James Nisbet. A Study for Young Men.
Extra fcap. 8vo, cloth, printed on toned paper,   -   -   0   2   6
———— Extra cloth, gilt edges,   -   -   -   0   3   0

Watts (Isaac, D.D.) Divine Songs for Children; with Scripture
Proofs. For the use of Families and Schools. Square 32mo, sewed,   0   0   2

White Ros of Glenmere (The).
By Mona B. Bickerstaffe. And other Tales. Extra fcap. 8vo, with
Illustrations, cloth, gilt edges,   -   -   -   0   3   0

Wilberforce (William). A Practical View of Christianity.
New and Complete Edition. Extra fcap. 8vo, cloth, -   -   0   2   6

Wise Sayings, and Stories to Explain them.
By M. H., Editor of "The Children's Hour." With Illustrations.

1. Jamie Wilson's Adventures.	7. The Nutting Party.
2. Little Silphy.	8. Castles in the Air.
3. The May Garland.	9. The Turnip Lantern.
4. The Visit to Eden Park.	10. The Lost Brooch.
5. The Two Friends.	11. The Village Favourite.
6. Alick Watson and his Quaker Friend.	12. The Gold Medal.

Done up in a neat Illuminated Packet. Super royal 32mo,   0   1   0
In cloth extra, gilt edges,   -   -   -   0   1   6

Witless Willie, the Idiot Boy.
By the Author of "Mary Matheson," etc. Royal 32mo, cloth,
bevelled boards, Illustrated,   -   -   -   0   1   0

CHORAL HARMONY,

FOR THE USE OF CHORAL SOCIETIES, SCHOOLS, ETC.

Price One Penny each Number.

———

Those numbers marked † contain easy Music for Elementary or School practice.
Those marked * have an Accompaniment.

SACRED.

3. O praise the Lord.		Colville.
6. Pray for the peace of Jerusalem.		
Hark, the loud triumphant strains. (12th Mass.)		Mozart.
†7. Brightest and best of the sons of the Morning. 3 v.		Webbe.
The Lord is my Shepherd.		Pleyel.
Be joyful in God.		Colville.
Characters used in Music.		
†8. Musical Signs and Abbreviations.		
How firm a foundation.		Mozart.
From Greenland's icy mountains.		Banister.
†11. To us a Child of hope is born.		Mason.
Hark, the herald angels.		Arnold.
Hallelujah.		R. A. Smith.
14. Make a joyful noise.		R. A. Smith.
Sanctus.		Camidge.
15. Sing unto God.		R. A. Smith.
17. Great God of Hosts!		Fowlie.
O God, forasmuch.		Fowlie.
*20. Blessed is he that considereth the poor.		R. A. Smith.
22. Hymn on Gratitude.		Holloway.
*24. Come unto Me.		
Now to Him who can uphold us.		R. A. Smith.
26. O Father, whose almighty power (Judas).		Handel.
*28. There is a land of pure delight.		Colville.
*31 & 32. The earth is the Lord's.		R. A. Smith.
*35. Jerusalem, my glorious home.		Mason.
*38. Hear those soothing sounds ascending,		Beethoven.
*39. Walk about Zion.		Bradbury.
He shall come down like rain.		Portogallo.
*43. Blessed are those servants.		J. J. S. Bird.
Enter not into judgment.		J. J. S. Bird.
*47. Ode on resignation.		Colville.
†48. Hark, the Vesper Hymn.		Russian.
The hour of prayer.		Douland.
Thanksgiving Anthem.		
God save the Queen.		
†50. God bless our native land.		

CHORAL HARMONY—*Continued.*

	Forgive, blest shade.								Callcott.
	Morning Prayer.								Herold.
51.	We come, in bright array (*Judas*).						Handel.
	Lead on, Lead on (*Judas*).							Handel.
†54.	Ye gates, lift up your heads.						Dr Thomson.
	O send Thy light forth and Thy truth,					R. A. Smith.
†55.	Who is a patriot.								
	Praise the Lord.								
	Gently, Lord, O gently lead us.						Spanish.
	Joy to the world.								
†59.	With songs and honours.							Haydn.
	Hymn of thanksgiving.							Mason.
	God is near thee.								
60.	But in the last days.							Mason.
*64.	Great is the Lord.							American.
	Arise, O Lord.								American.
*69.	Awake, awake, put on thy strength, O Zion!						
*70.	I will bless the Lord at all times.					R. A. Smith.
*71.	Hallelujah! The Lord God omnipotent reigneth.				R. A. Smith.
	God, the omnipotent.					Russian National Melody.
†72.	The brave man.							H. G. Nageli.
	Lift up, O earth!							G. F. Root.
	From all that dwell below the skies.						
	When shall we meet again.								
	O wake, and let your songs resound.					Himmel.
	All hail the power of Jesus' name.						
*75.	Blessed be the Lord.							R. A. Smith.
	Great and marvellous.							R. A. Smith.
*77.	Grant, we beseech Thee.							Callcott.
	Come unto Me when shadows darkly gather.					
79.	The Lord is my Shepherd.						Beethoven.
	Let Songs of endless praise.						Mason.
	My Faith looks up to Thee.						Mason.
81.	Beyond the glittering starry sky.					J. Husband.
82.	Blest Jesus, gracious Saviour.					Michael Haydn.
	Hymn of Eve.								Dr Arne.
	Salvation to our God.								
84.	I will arise.								Richard Cecil.
	Blessed are the people.								
86.	I was glad when they said unto me.					Dr Callcott.
87.	Hark! above us on the mountain.					C, Kreutzer.
88.	Then round about the starry throne.					Handel.
*91.	Oh! how beautiful thy garments, O Zion.				J. A. Naumann.
*92.	Put on thy strength, O Zion.					J. A. Naumann.
*98.	Sing to the Lord, our King and Maker.				Haydn.

The Series to be continued.

CHORAL HARMONY—*Continued.*

SECULAR.

1. Let no dark'ning cloud annoy.	*German.*
The Reapers.	*Colville.*
2. There is a Ladye sweet and kind.	*Ford.*
Gentle Spring.	*Colville.*
4. And now we say to all, Good-night.	*Methfessel.*
The fountain.	*Colville.*
5. Good Morning.	*Bradbury.*
Swiftly, swiftly, glide we along.	*Colville.*
†9. May-Day. *Colville.*—Harvest time.	*Storace.*
Glossary of musical terms.	
†10. Spring-time. *Silcher.*—Freedom.	*Scottish.*
Rosy May. *Scottish.*—The Daisies.	*Mozart.*
†12. Summer's Call.	*Colville.*
Midnight.	*Donizetti.*
13. Hark, the Curfew's solemn sound. 3 v.	*Attwood.*
16. Serene and mild.	*Webbe.*
18. How sweet, how fresh this vernal day.	*Paxton.*
Stars of the summer night.	*Cokking.*
19. Thyrsis, when he left me.	*Callcott.*
21. The Coquette. The Exquisite.	*Neithardt.*
Aldiborontiphoscophornio. 3 v.	*Callcott.*
23. Swiftly from the mountain's brow.	*Webbe.*
*25. It is better to laugh than be sighing.	*Donizetti.*
27. Hark, the hollow wood surrounding.	*J. S. Smith.*
It was an English Ladye bright.	*Hine.*
†29. Joyful be. *Schneider.*—Sweet peace.	*K. Smith.*
O lady fair. The last rose of summer.	*Moore.*
30. The Skylark's song.	*Mendelssohn.*
Spring Morning.	*Schneider.*
†33. Come and join our trusty circle.	*Gabler.*
The Forest. *Karew.*—Sweet love loves May.	*Silcher.*
*34. Glad May-day.	*Neithardt.*
36. Good-night.	*Hulme.*
Bright, bubbling fountain.	*Waelrent.*
37. From Oberon, in fairyland.	*Stevens.*
*38. The Chapel.	*Kreutzer.*
†40. 'Tis dawn, the Lark is singing.	*G. Webb.*
Thrice hail, happy day.	*German.*
Home! Home!	*Paz.*
Come joy, with merry roundelay.	*German.*
41. Sweet Echo, sweetest nymph.	*Birch.*
*42. The Gleaners.	*Mendelssohn.*
*44. The Sight Singers.	*Martini.*
Hail, festal day.	*Rossini.*
45. Thy voice, O Harmony.	*Webbe.*
46. Rural pleasure.	*Kreutzer.*

CHORAL HARMONY—*Continued.*

See the Sun's first gleam.	*Schuffer,*
49. The Sprite Queen.	
The Sun's gay beam.	*Weber.*
Behold the morning gleaming.	*Weber.*
52 & 53. All the Choruses usually performed in Locke's Music for Macbeth.	
55. Hail, smiling morn.	*Spofforth.*
See our oars with feather'd spray.	*Stevenson.*
57. Come, gentle Spring.	*Haydn.*
†58. Never forget the dear ones. 3 v.	*Root.*
Merrily o'er the waves we go.	*Bradbury.*
The foot Traveller.	*Abt.*
61. The Chough and Crow. 3 v.	*Bishop.*
62. The huge globe has enough to do. 3 v.	*Bishop.*
63. May Morning.	*Flotow.*
Come to the woody dell.	*Pelton.*
65. Which is the properest day to sing?	*Arne.*
Beat high, ye hearts.	*Kreutzer.*
66. Now strike the strings.	*Rudd.*
Since first I saw your face.	*Ford.*
67. Step together.	*Irish Melody.*
For Freedom, Honour, and Native Land.	*Werner.*
The Mountaineer.	*Tyrolese Melody.*
What delight, what joy rebounds.	*German.*
68. Come, let us all a Maying go.	*L. Atterbury.*
Hark, the Lark.	*Dr Cooke.*
Here in cool grot.	*Lord Mornington.*
*73. Come on the light winged gale. 3 or 4 v.	*Callcott.*
*74. Sleep, gentle Lady.	*Bishop.*
76. Sparkling little fountain.	*Bradbury.*
The dazzling air.	*Evans.*
*78. On Christmas Eve the bells were rung. 3 or 4 v.	*P. King.*
80. Hail, all hail, thou merry Month of May.	*G. Shinn.*
83. The sea, the sea, the open sea.	*C. S. Neukomm.*
85. The Singers.	*C. Kreutzer.*
89. Call John.	*American.*
The Travellers.	*James King.*
90. Laughing Chorus.	*G. F. Root.*
Soldier's Love.	*F. Kucken.*
*93. Foresters sound the cheerful horn.	*Sir H. R. Bishop.*
*94. Gaily launch and lightly row.	*Mercadante.*
My lady is as fair as fine.	*John Bennett.*
*95. See the bright, the rosy morning.	*Carl Blum.*
The land of the true and brave.	*Franz Abt.*
*96. What shall he have that killed the deer?	*Sir H. R. Bishop.*
*97. The Song of the New Year.	*Donizetti.*
*99. Why should a sigh escape us?	*Otto.*
How sweet the joy.	*C. Kreutzer.*
*100. Upon the Poplar bough.	*Paxton.*
Mountain Home.	*C. Kreutzer.*
Over the Summer sea.	*Verdi*

www.ingramcontent.com/pod-product-compliance
Lightning Source LLC
Chambersburg PA
CBHW030632030726
47497CB00006B/1755